James Rodway

In the Guiana Forest

Studies of Nature in Relation to the Struggle for Life

James Rodway
In the Guiana Forest
Studies of Nature in Relation to the Struggle for Life

ISBN/EAN: 9783337026660

Printed in Europe, USA, Canada, Australia, Japan

Cover: Foto ©berggeist007 / pixelio.de

More available books at **www.hansebooks.com**

IN THE
GUIANA FOREST

*STUDIES OF NATURE IN RELATION
TO THE STRUGGLE FOR LIFE*

BY

JAMES RODWAY, F.L.

Author of
"A HISTORY OF BRITISH GUIANA," ETC.

WITH INTRODUCTION BY GRANT ALLEN

ILLUSTRATED

LONDON
T. FISHER UNWIN
PATERNOSTER SQUARE
1894

CONTENTS.

CHAP.		PAGE
I.	THE FOREST	1
II.	THE MAN OF THE FOREST	17
III.	THE ANIMALS OF THE FOREST	44
IV.	INTERDEPENDENCE OF PLANTS AND ANIMALS	62
V.	THE STRUGGLE FOR LIFE	82
VI.	ON THE RIVERS AND CREEKS	101
VII.	UP IN THE TREES	121
VIII.	IN THE SWAMP	141
IX.	ON THE SAND-REEF AND MOUNTAIN	158
X.	ON THE SEA-SHORE	169
XI.	IN THE TROPICAL GARDEN	187
XII.	MAN'S FOOTPRINTS	199
XIII.	THE SENSES OF PLANTS	211
XIV.	THE CAUSES OF THE STRUGGLE	224

LIST OF ILLUSTRATIONS.

FACING PAGE

1. THIN TIMBER ON POOR SOIL . *Frontispiece*
2. SECOND-GROWTH FOREST 2
3. CLUSIA FLOURISHING ON LOCAL TREE . 90
4. CONFUSED ASSEMBLAGE OF AËRIAL ROOTS OF A CLUSIA 92
5. WILD-FIG, AFTER DESTROYING ITS HOST . 94
6. CREEK SCENE 100
7. SILK COTTON-TREE, CROWDED WITH EPIPHYTES 122
8. AN ORCHID, ASSOCIATED WITH OTHER EPIPHYTES 130
9. INUNDATED FOREST . . . 154
10. ON THE LAND REEF 158
11. ABORTIVE AËRIAL ROOTS OF THE COURIDA . 170
12. AËRIAL ROOTS OF COURIDA GROWING UPWARDS 172
13. MATTED ROOTS OF COURIDA . . . 174
14. IN A MANGO SWAMP 182
15. A JERMINALIA OVER-RUN BY LORANTHS . . 188
16. AN INDIAN PATH . . . 200

INTRODUCTION

THE tropics, I have often said, are biological head-quarters.

It is not merely that there, and there alone, do you see life at its fullest, its fiercest, and its fieriest. It is not merely that there do you find the struggle for existence carried on with a wild energy which none can overlook, both among plants and animals. It is not merely that there trees, shrubs, and herbs, beasts, birds, and reptiles abound, with a richness and a variety unknown in more temperate regions. The tropics have a far deeper value for the biologist than all that. They are typical and central. They are ancient and historical. They represent in our cooled and degenerate world, from which all the most Titanic forms have fled, the circumstances under which plant and animal life first arose, and under which it passed through the main stages of its evolutionary history. This it is that gives value to the tangled brake of tropical scenery in the eyes of the biologist; he sees in that crowded and bustling woodland the image of the great world where our forefathers were nurtured.

Europe is a creation of the Glacial Epoch. The Great Ice Age, that nightmare of science, swept like a gigantic plane over our northern latitudes, and killed out at one blow all our largest, fiercest, and most interesting animals. It came upon a world of colossal species. When it cleared away for ever, the mammoth was gone; the sabre-toothed lion had disappeared; the hippopotamus and rhinoceros had retired in disdain to the banks of the Nile or the jungles of the Ganges. A cold and stunted fauna and flora replaced the wild growths of the Eocene and Miocene. Life faltered and failed. The birch took the place of the palm and the custard-apple; where magnolias and cinnamons once reared their stately heads, the holly and the bramble, the gorse and the white-thorn overran our lowlands. It is from this impoverished northern world, the leavings of the Ice Age, that we Europeans and New Englanders of to-day form our first conceptions of the life of ages; it is from our study of these chilled and depauperated sub-arctic forms that we strive for the most part to reproduce in our mind's eye the fierce and teeming panorama of the evolution of species.

For a truer and deeper picture of the beginning of life, the veritable theatre of the origin of species, we must go to the tropics. There, the conditions which existed from the beginning upon our warmer planet have been more faithfully continued to the present day : there, the traditions are more unbroken, the history of vegetable and animal life more realisable. True, even there, the Glacial Epoch played havoc with the leaders in the race; the effects of that great secular cooling, though far less severe than in more northern and more southern climes, were sufficiently disastrous to mark the end of a period. Life is nowhere what it was in the palmy days of the deinotherium and the mastodon. But in the tropics, it gives us, at any rate, some faint idea of that luxuriant world in which the fierce battle of the kinds was first fought out, and in which, as I at least believe, the early ancestors of man first began to be fairly human. Those mighty forests, with their strangling creepers, enable us to picture to ourselves, however dimly, the scene that unfolded itself before the evolving eyes of all evolving nature in its greatest epoch.

For this reason, I have always felt deeply impressed by the educational value of a sojourn in the tropics, not merely for the biologist, but also and perhaps almost equally for the general thinker. The tropics are nature. Certainly it is most noteworthy, as I have pointed out elsewhere, and as Mr. Rodway again points out in this interesting volume, that the very idea of evolution and of the survival of the fittest is to a great extent a direct product of tropical travel. It occurred to Darwin himself during his tropical experiences in South America, while he was voyaging in the *Beagle;* it occurred independently to Alfred Russel Wallace amid the shady depths of the Malayan forests. Bates had half anticipated it on the banks of the Amazons; Huxley had stored up facts for it in his coasting trips along the northern shore of tropical Australia. Its later prophets, like Belt and Fritz Müller, were tropical travellers to a man: it was only the stay-at-home Owens and Virchows, the laboratory naturalists, who sturdily opposed it. Thus the value of a tropical education as biological training is immediately proved; its value as a general widener of social and economic views, in like manner, can hardly be overrated.

Now the best thing, of course, to show us what the tropics are like, is to go and see them. Failing that, the next best thing is to be personally conducted round that great estate by a thoroughly competent and observant showman, in imagination, on printed paper. For this purpose Mr. Rodway is one of the best guides I have ever come across. His book speaks for itself; nevertheless, I will venture also to speak for it. Though good wine needs no bush, a good book is sometimes all the better for a few passing words of friendly introduction. The author of these pretty and interesting essays is fortunate in superadding to the eye that sees the tongue that speaks, well aided by the pen of a ready writer. Furthermore, he has the invaluable gift of picturesqueness. He can call up better than almost any one I know the exact tone and spirit of those solemn silences, those suggestive glooms, that brood eternally over the mystic soil of the tropical forest. My own equatorial experiences have been mainly made in outliers of the same rich regions which Mr. Rodway describes; and I can therefore testify to the essential fidelity—a fidelity not only of the letter but of the soul as well—which marks his descriptions of those mighty woods—natural rivals of Karnak or Denderah—where huge columns rise buttressed to the sky from bare forest glades, supporting one vast dome of living green, through which scarce a ray of subdued light flickers timidly down to the leafless and flowerless bed of leaf-mould beneath them. In that wonderful world our author is at home. He has threaded those strange rivers of tangled vegetation in his native boat with his Indian guide to no small purpose; and he reveals for us here all the results of his experience with a justness of language and a vividness of style which fairly entitle him to be called a Jefferies of the tropics.

Mr. Rodway is fortunate, too, in his choice of a locality. No tropical country is more interesting than South America. It is true, the hotter parts of Asia, and even of Africa, can give us in some ways a more varied, a more advanced, and a more colossal fauna. They represent more closely the conditions under which the leading types of life in the great continents were first evolved, and they supply us with the nearest existing prototypes of our anthropoid ancestors. Compared with these, the fauna of South America is small in size and antiquated in pattern. But it is just this in part which gives it its greatest charm. Australia, to be sure, is more antiquated still; but then, the mammalian fauna of the Australian continent antedates the evolution of the placental type—it consists entirely of marsupials and monotremes. Australia is still, as it were, in its cretaceous period. This deprives it of the historical interest, for our race at least, which attaches to the earlier stages in the evolution of the dominant placental mammals and of our own immediate sub-human ancestry. Now South America in these respects holds an intermediate place between very antique Australia and very modern Asia; it preserves for us, as in amber, numberless intermediate stages, younger in order than the Australian kangaroos and wombats and phalangers, but older than the lions, the elephants, and the

Asiatic or African monkeys. It is a halfway house in the history of evolution, and possesses for many of us, therefore, that indefinite charm which the Middle Ages possess as a half-way house between Græco-Roman civilisation and the squalid modern industrial system.

Why is this? Mr. Wallace has answered that question for us beforehand. He has shown us that South America remained for many ages a great island continent, like Australia at the present day, cut off from direct intercourse with the world to the north of it, and thrown entirely on its own resources. As yet there was no Panama to vex the soul of a de Lesseps; the animals, and to a less degree the herbs and trees of this isolated mainland developed slowly along lines of their own, in adaptation to the special peculiarities of their remote habitat, without danger of influx from the larger lands, where the struggle for life was producing meanwhile very different results on the same original ancestral models. South America, in short, was a biological Japan to the tertiary Europe. Much the same thing happened also with South Africa, where a local fauna and flora still hold out to a great extent against the intrusive types which have invaded it from the Asiatic and South European region. But in tropical America the isolation continued for a much longer period; Panama, as Mr. Wallace has shown reason to believe, is but a thing of yesterday. Moreover, the great American island was larger than the South African one. Hence, it was enabled to develop a very varied and extensive fauna of its own low types; while, owing to the narrowness and mountainous nature of the connecting isthmus, when it *did* arise, Brazil and Guiana have never been flooded and overrun by intrusive northern genera and species to anything like the same extent as has happened in South Africa. South America, in other words, still preserves for us to-day a number of comparatively early and low types, just a stage higher than the Australian level, but inferior to the general facies of the fauna in the great northern continents.

Edentates, marsupials, and rodents form the larger element of the mammalian fauna in the Neotropical region. These include a considerable number of opossums, less advanced in structure than the North American species; as well as the sloths, the armadillos, and the true ant-eaters. The elephants and rhinoceroses are altogether wanting; in their place we find only that earlier and lower form, the undifferentiated tapir, elsewhere extinct, save in the Malayan region. The pigs once more, are only represented by the diminutive peccary. The camels, an antiquated type even in the old world, find their analogue in the still more antique and unspecialised llamas, alpacas, and guanacos. Sheep, oxen, and antelopes are entirely absent. The monkeys all belong to inferior types, marked by the broad lateral nostril; and most of them have the peculiar prehensile tail which goes with a poor development of the opposable thumb. Among their allies, the marmosets preserve for us a stage in the early evolution of the race not elsewhere represented. On the other hand, the greater cats have made good their foothold, and are well vouched for by the puma and the jaguar.

Among birds, somewhat analogous facts may be noted. The Picariæ, for example, greatly outnumber the Passeres; and the latter, which include in other regions all the true song-birds, show their early character in South America by being mostly songless, and with the vocal muscles absent or very feebly indicated.

The features I have thus sketched in rapid outline belong, of course, to the continent as a whole. Of this vast kingdom, Mr. Rodway has tilled but a tiny province. Within its limits, however, he has done good service, and has rendered an admirable account of his talent. What dominates his book is the informing sense of the struggle for life as the condition precedent of natural selection. In this respect, he reminds one of Belt in Central America. His book gives the untravelled reader a far better conception of the lonely green gloom of South American forests, and the tragedies they cover, than many more pretentious works of more ambitious explorers.

Not the least interesting portion of this volume to my mind, however, is that which deals with the haunts and ways of the human denizen of the forest. It is written with discrimination, with insight, and with sympathy. And, what is rarer still, it does justice to the savage. "The only good Indian," said General Custer, in an oft-quoted epigram, "is a dead Indian"; and Artemus Ward echoed the General's sentiments in his famous dictum, "Injins is pison wherever found." They were thinking of the debased and degraded Indian of the North, the product of two centuries of degenerating contact with American civilisation and American whiskey. It would have been fairer to say, I think, that bad white men make bad Indians. Fifty years of Sioux massacres are likely to produce a certain impatience of the ways of the pale-faces on the part of the redskin. The Indian of the South American forest, on the other hand, is a very different creature, a true denizen of the woods, unspoilt as yet by alcoholic civilisation, and undecimated by the free use of Martini-Henrys. Mr. Rodway has described him in a charming essay with rare pictorial skill, showing here as elsewhere a power of conveying his own impressions to others which is of a high order of literary merit.

Last of all, it is too often the fault of tropical travellers that they care a great deal for skins and horns and feathers, but little or nothing for leaves, fruits, and flowers. This is a sad mistake, for the vast majority of living things, after all, are plants or trees, while the animals are everywhere in a miserable minority. This error Mr. Rodway, whom I take to be above all things essentially a botanist, has successfully avoided. It is the woodland he paints, not merely the sport in it. He gives us a faithful picture of that ever-present reality, the forest itself, as well as of the beasts, and birds, and reptiles, and insects that haunt it or burrow in it. This ever-present reality, which too often becomes a mere background in the picture to the animal life, is here represented of its proper size and in its appropriate proportions. It *is* the portrait. This peculiar touch, I think, makes Mr. Rodway's book one of the most impressive and weirdly solemn delineations ever limned by cunning hands of the great tropical woodlands.

GRANT ALLEN.

IN THE GUIANA FOREST.

I.

THE FOREST.

EVERY one loves the forest. Whether it be the merry green wood of the bold outlaw, the pleasant grove, or the game preserve, each has its charm. If such is the case with the insignificant assemblages of trees in temperate climates, what shall we say to the great tropical forests. As we sit down and recall the memories of different excursions, we are almost carried away with enthusiasm and wish to be once more lying in our hammock on the bank of some grand river. Visions of mighty streams with dangerous rapids, of creeks winding amid flowery arcades, of wide expanses of savannah and swamp bordered by lines of palms, of sand-reefs glittering in the intense sunlight, rise up before us and present pictures of loveliness that makes us feel how dear such memories are. No matter that every trip was more or less uncomfortable, and some even dangerous, our pulse beats excitedly as we think of nature in all her wildness and beauty. We may spend days or weeks in an open boat, now exposed to the intense glare of a tropical sun, and anon drenched by sudden downpours of rain of which other climes know nothing, yet our memories remain as dear to us as if everything had been pleasant. By day the journey was long and tiresome, and when night came our rest was perhaps disturbed by mosquitoes or sand-flies, yet the magnificent scenery of the forest has been impressed upon us in a way never to be forgotten. Every trouble and difficulty—every danger to health or life—has been gilded by a covering of beautiful leaves and flowers.

This enthusiasm for life in the bush is common to all, but more especially to the huntsman and naturalist. An ordinary traveller, however, soon gets wearied of the awful silence; to him everything is uniform—the forest is simply a jungle or wilderness. He sees before him a confused assemblage of pillars, supporting an enormous green roof, of which he can distinguish but little

SECOND GROWTH FOREST.

on account of the dense shade. Then, the heat and moisture, together with the oppressive smell of decaying vegetation, remind him that fever lurks in these recesses, and drive him back to his bateau. However, indifferent as he may be to the primitive forest, he cannot but enjoy a trip on one of the smaller rivers or creeks. Here each bend brings a new picture before his eyes, every one worth keeping in mind until his return, when, perhaps surrounded with home comforts, he will forget all the inconveniences which prevented his enjoying them at the time.

To the huntsman, with whom we must class the native Indian, the forest is never dull. However deserted it may appear, he knows where to look for the acourie and labba, or on what trees monkeys and parrots are likely to be found. The Indian treads softly so that not a twig or leaf rustles under foot, yet he trips along much faster than the white man can follow. To him the pathless woods are familiar; he knows every hill and valley; the little streams which trickle downward have their destinations, and he can tell you whence they come and where they go. Then, hundreds of the great trees are quite as well known to him as are public buildings to the inhabitants of the city. If in a strange locality he bends down a young sapling at intervals, so that the under surface of the leaves may be visible on his return and point the way. His white companion misses him as he hurries along and stands up bewildered. At a short distance his reddish-brown skin harmonises with the tree-trunks, and as his movements are absolutely noiseless, he cannot be seen or heard. Perhaps the Indian wants to shoot some animal, and purposely leaves the white man behind for fear he should disturb the game. Then the report of a gun is heard reverberating among the columns, but even that will not indicate the direction to any one but an expert. Presently the Indian comes back with his game and smiles at the remonstrances of his companion. On they go again until perhaps a small creek is reached, over which stretches a fallen palm. On this the red man lightly trips and is going straight along into the forest, expecting the other to follow. But who could walk in boots on such a slippery bridge? The white man gives the bush call, Hoo—oo—oo—oo, and the Indian comes back again looking surprised that this convenient bridge could be any more an obstacle to another than to him. The novice tries to explain, but almost before he can understand what is taking place, he is in the arms of his companion and over the creek.

The Indian is at home in the forest. His colour is in harmony with his surroundings and his hut seems altogether fitted for the little mound on which it generally stands. Its palm thatch soon becomes brown and differs little from the dead leaves which are falling in every direction, while its uprights are only small tree stems which can be matched a few yards away. No painted walls or carving of any kind obstruct the view, nor do his few articles of furniture disturb the general effect. When we see a boarded cottage on the bank of the river with its surroundings of stumps, and perhaps charred logs, a contrast is at once produced, and the gap is a blot on the landscape, but when an Indian makes a similar clearing for his cassava field he always chooses a place at some distance from the settlement, where, surrounded by tall trees, it can only be found after a careful search.

As the home of the red man harmonises with the forest, so does his canoe with the dark waters of the creek. He uses no ornament on his frail craft, so that whether it lies on the surface of the water, is hidden in a tangle of bush-ropes, or drawn up on the bank, it is hardly distinguishable from its surroundings. As he sits and paddles quietly along it makes us think of a possible explanation of the old myth of a creature half-man, half-fish. In the gloom of the forest arcade the canoe glides along as noiselessly as if it were a brown aquatic creature with a man's body and shoulders above water slowly guiding itself with one great flapper. This is best seen in small canoes, some of which hold only one person. They are so very unstable that none but Indians dare to use them even in the stillest water. Yet their owners will take them into the breakers below the rapids, and even stand up in them to shoot the great pacou and other fish which abound in such places.

If the forest has attractions for the huntsman, how much more interesting it must be to the naturalist. What one who has delighted the world for over fifty years thought of the Guiana forest may be seen in Waterton's "Wanderings." The enthusiasm of the Yorkshire squire has probably never been surpassed. To him the forest was something more than the awful solitude which is the first impression it makes on a stranger—it was full of life. No matter that the silence at noonday was something almost appalling and that the unfamiliar cries of animals after sunset only added to the impression of loneliness, a feeling of intense but quiet pleasure was produced which can be appreciated only by one who has followed in his steps.

Unfortunately for Charles Waterton he was not a botanist, and therefore could appreciate only one side of forest life, and that possibly the least important. More and more every day it becomes necessary for the naturalist to go beyond his own special province. The entomologist must know something of the plants on which insects feed, and the botanist a great deal about the fertilising agents continually at work among the flowers. He who sees only one aspect of nature can never fully appreciate the beautiful adaptations of one to another and their perfect interdependence.

Naturalists have been stigmatised as wanting in that sense of beauty and harmony so common everywhere among poets and dreamers. Far from such being the case, however, none but a student of nature can fully appreciate a landscape. The painter sees patches of colour in sky, tree, or river, but the naturalist recognises the objects which make up the scene. On the sand-reef he distinguishes the footsteps of a jaguar and the remains of his dinner, and can picture what has taken place in the night. A peccary left her hole in a hollow tree at nightfall to feed under the saouari-nut trees. She is quietly cracking the shells and munching the oily kernels, when the great cat suddenly pounces upon her, and notwithstanding her struggles, of which the evidence is plainly marked on the sand, she is torn to pieces and eaten. Here are the delicate hoofs and there two or three bones, while bristles like miniature porcupine quills litter the sand.

Sitting on a hollow tree beside a creek, he sees a thousand flowers and fruits floating down the stream. Now he distinguishes a palm nut snatched under the water by a great fish, or a shoal of small fry feeding on the yellow hog-plums which are so conspicuous against the dark water. Now there is a splash as an alligator comes out of the thicket and dives under, to come up again some distance away, hardly distinguishable except to a trained eye. This reminds us of the protective coloration of every living thing in the forest. The jaguar lives on the sand-reef where bushes grow in large clumps between irregular patches of sand. Unlike the dense forest, where reigns eternal twilight, these shrubs admit a few rays through the canopy above which lie as bright spots on the litter of dead leaves. How like is this to the markings of the jaguar, and how easy can this beautiful creature lie hid in such a thicket. Again, the tapir and a species of deer have white markings when young, which they lose as they grow older; these are also protective at the time when such protection is most necessary. Other protective contrivances are found in every forest animal, the sloth being especially remarkable for its long hairy coat and its manner of hanging under a branch like one of those nests of termites so common in the forest, or the cluster of aerial roots of an epiphyte. Snakes are also nearly invisible in the gloom, notwithstanding their brilliant colours when played upon by the sunlight.

When the sun is high, from about eight o'clock in the morning until four in the afternoon, hardly a sound is heard. Every animal is asleep or quiet, and not even a bird utters its characteristic note. Now and again, however, a howling monkey or a tree-frog breaks the silence with his prolonged notes, but these only add to a feeling of utter solitude, as does the baa of a sheep on some mountain pasture in other climes. Then, the forest is steaming with moisture like a Turkish bath, and man feels inclined to lie in his hammock and take a siesta. The open river glows with the fervent heat, the surface is warm to the touch and even the fishes retire into the depths below, so that not a ripple disturbs the surface. Near the banks every bush and tree is mirrored on the smooth surface, and among these shadows a canoe may sometimes be distinguished moving noiselessly along as if it were a phantom.

With so few atmospheric changes it might be supposed that the tropical forest would give rise to little variation in animals and plants, yet, on the contrary, it is here that nature runs riot as it were. Physical environment has certainly some influence, but we must look for the prime factor of differentiation in the intense struggle for life. Man has done little or nothing towards producing the result, but everything is the outcome of energy and selfishness. Nature has been lavish with her gifts. Heat, light, and moisture have been plenteously bestowed, yet few trees can get room to assimilate as much of the two latter as they need. The forest is densely populated—more so, in fact, than any city ever was or could be. There is not room for one in a thousand of the children born therein, so that the fight for standing room is like that of a crowd at a *fête*. It follows, therefore, that every possible contrivance to gain a position has been developed, and the result is almost perfection. The victory, however, has not been to the few, but to the many. One species has progressed on certain lines, but others have not been idle; although the developments may be different in each case the end is almost identical.

The trees have succeeded in defending themselves against almost every animal. It is true that the sloth, and certain species of ants and caterpillars, almost strip them of their leaves, but not a single animal appears to gnaw at their bark. Even the young seedlings are free from outside enemies, and have every possible opportunity of gaining a position if their elders give them a chance. The fight is therefore one between tree and tree; not even species, but individuals. Like a battle of the Middle Ages, the fight is made up of single combats, where each forest giant is a centre with enemies in every direction. And, not only is it a battle, but a fight for life which has to be continued day after day and year after year without cessation. There never was, nor ever can be, a truce. The armour cannot be put off, but must be continually worn and always kept in order. Now and again one of the combatants is killed, and then ensues a struggle for his position. There is room for one, and a hundred little soldiers have to fight until the fortunate victor gains the place, and can take its part in the greater struggle.

"Practice makes perfect." An old soldier necessarily becomes proficient in warfare, and under the system in vogue during the Middle Ages the result generally depended on individual prowess. Such is the case now in the Guiana forest. Every tree has chosen its weapons of offence and defence, and taught itself how to wield them to the best advantage. No species is armed in exactly the same way as another, but every one made his selection in some past age, and goes on year after year making improvements as his enemies become stronger. Why one should choose to strangle his enemy, another to suck his blood, and a third to smother him, it is hard to say, but such differences exist.

How firm and stately are these great monarchs of the forest! They have gained their positions years ago, and are now strong enough to hold their own. They have good armour, well tested, have killed off thousands of younger rivals, and might be thought secure. But no, mere strength may not overcome them, but craft will. Like Sinbad, they take some apparent weakling on their shoulders, and he soon develops into another "Old Man of the Sea." When the forest giant gets old and feeble his enemies come round him in force. He can no longer continue the great struggle, and quickly succumbs, perhaps to two or three of his own children who have grown up under his shadow.

In the midst of such a fight it naturally follows that every generation becomes stronger and more fit for the struggle. Every experience of the parent is perpetuated in the child, so that it is easily conceivable that the son may overcome his own father by means of higher powers gained in the lifetime of that individual, but which were less pronounced in the former generation. When we look upon every individual man, beast, or plant, as nothing less than the same being who lived tens of thousands (perhaps millions) of years ago, and has continued to live up to the present moment, what wide views of life we acquire. We are our parents, grandparents, and ancestors of all past ages, up to that simple cell which first showed a germ of life. These were not only our relations, but our very selves, and as long as the chain continues we have endless life.

We cannot affirm that the Hebrews held this theory of continuity, but many indications of something like it are nevertheless found in their history. Nature abhors a creature without offspring, and from her standpoint the man who does not procreate children is at least unnatural, if nothing worse. Every tree of the forest, from youth to old age, is doing its best to ensure that his line shall not come to an end with him. In youth he does everything possible to become strong enough to fight for a place among others. For what purpose? That he may get sunlight to produce offspring. If he is overshadowed by others, flowers cannot be produced, and without flowers, as a matter of course, there can be no seeds. When we see the plants of our gardens flower freely we know they are in good health. We can even conceive that they glory in the success they have achieved. How much more, then, must the trees of the forest exult in the fact that they have at last overcome all opposition and produced their first-born.

Seeing all these things and a thousand others in the forest, the naturalist cannot but feel interested in it. Every day something new attracts his attention, and every fresh observation gives rise to a train of thought which leads him on and on to solve problems of the greatest importance. At first the number of these is appalling—he gets almost disheartened as questions come up one after another. To slightly alter an old quotation, "*Science* is long but life is short," nevertheless, if we succeed in discovering a few of nature's secrets, we shall not have lived in vain. Unfortunately for tropical nature, few naturalists have had the privilege of even a passing glimpse of her beauties, while those who have lived in the great South American forest for any length of time could be numbered on the fingers.

No doubt the investigation of tropical nature is responsible for the evolution theory. Even if this be questionable in Darwin's case, although it seems evident that his voyage in the *Beagle* gave him the first glimmerings of light on the subject, it cannot be disputed in that of Wallace. Then Bates, Belt, and a number of others who have collected evidence in support of the theory have all lived in the tropics and forwarded their observations to be dealt with by students all over the world. Here, in the Guiana forest, the evidence for continual development is beyond question. Every tree, every animal, and every man is a living example of heredity and environment—natural and sexual selection. The struggle for life and the survival of the fittest can be seen every day in the forest, on the banks of the rivers, among the sedges and grasses of the swamp, along our muddy shores and in our semi-wild gardens. The man of the forest, when untainted by contact with his more civilised cousin, also, like every other animal, shows that evolution is at work to-day as it has been through all past ages. Every living thing is ever moving forward, working towards an end which is unattainable—perfection. But, although this object will never be achieved, the results of the struggle bring it continually a little nearer, and therefore cannot be otherwise than good. No matter that they never reach a point where conflict is no longer necessary, improvement is visible in the survivors. Even the extinction of a particular family or species makes room for others more fit to carry on the strife. Nature does not take care of the weaklings, she provides no asylums, if some of her creatures cannot work for a living they must make room for those who can. Individuals are of little consequence as such, but nevertheless as links in the great chain they are of the greatest importance.

II.

THE MAN OF THE FOREST.

THE man of the forest is in almost perfect harmony with his surroundings, and if to be so is to be happy, as some have said, then the South American Indian must be one of the happiest of men. He certainly is a most pleasant fellow, and if unsophisticated nearly always kind and obliging. Without him the traveller could hardly find his way through the trackless forest, while in his company everything is comparatively easy. Like other men, as well as animals and plants, he is uncomfortable away from his environment, and is looked upon by the low-class white, or even negro, as vastly inferior. If to be a savage means to be rude and uncouth, ill-mannered and disagreeable, then the Indian little deserves such an appellation. He is one of nature's gentlemen, and even when his wishes do not correspond with yours his opposition is only passive.

In British Guiana, where there are representatives of almost every race in the world, it is interesting to note the most striking differences between them. The European, and in a lesser degree the Hindoo, seems to have almost perfect control over his feelings. He can rule these if he wish, but when they do come to the front it is easily seen that they are very strong. The negro, on the contrary, carries his passions on the surface, and appears to have little control over them; it is even doubtful whether he appreciates the deeper feelings of humanity at all. The Chinaman, again, must have very strong emotions, but these are so completely hidden that it is almost impossible to discover any outward signs of their existence. Finally, the American Indian resembles the Chinaman in the almost total suppression of all evidence of deep feeling, and this is so much the case that we might be inclined to think him incapable of thinking. No doubt many of his duties are performed instinctively, but still there are signs that he can be roused from his apathy when occasion requires.

Having lived in the forest for ages, the Indian can hardly be looked upon as one of the rulers of creation, but rather as in perfect unison with nature. He is as much a part of the whole as the jaguar, the howling monkey, or the tapir. He does not interfere with the constitution of things—does not clear great tracts of land—builds no cities—erects no monuments—nor does he leave many more traces of his presence than the other inhabitants of the forest. His settlements have been scattered over the length and breadth of the land for ages, but except for a few trails, hardly more distinct than the runs of large animals, there are few signs of his presence. From one point of view he may be considered as having attained perfection. The balance of life has been kept up, and, apart from outside influences, he does not exterminate a single animal. Nowhere perhaps is the fauna of such an ancient type, so well protected, and so perfectly fitted to its environment, and nowhere can we study man as an animal so well as in the Guiana forest.

Whether it is a good thing for the Indian that he has accommodated himself so perfectly to his surroundings is doubtful. Apart from outside influences he would go on for ages, making little progress, and hardly increasing in numbers. The land could never be densely populated, as the number of inhabitants depends on the amount of food, and this again depends on the number of huntsmen. The man of the woods keeps no domestic animals for food—as for his bread (cassava), it could hardly support life without meat. He is, therefore, almost as much a beast of prey as the jaguar. Every day fishing and hunting excursions have to be made, and the experiences of all past generations empower to track the deer, tapir, and labba in such a way that he is nearly always successful.

Like the wild animals the Indian retires before the white man. He cannot accommodate himself to a new environment, and therefore moves on. Where, as in the West India Islands, he was prevented from migrating, he became extinct. Unlike the negro, who appears to accommodate himself to even great changes, the Indian dies when his environment is altered. The histories of the Spanish conquest give most horrid details of what was called by the new-comers obstinacy. Rather than labour on plantations or in mines, great numbers committed suicide, while others were flogged to death or executed, because they would not submit. To die rather than become a slave shows perhaps greater nobility than does abject submission, but it is not so conducive to perpetuation of the race. Savage communities generally consist of two classes, masters and slaves, and every man, woman, and child must belong to one of these. The South American Indian, however, has never adopted such a system. In early times there was a kind of association for mutual defence, under a war chief, elected for the purpose, but this was only a temporary arrangement. The chieftain was not any the less bound to hunt and fish for his wife or family—his benab was not a palace—nor had he any servants. We read of Indian slaves among the Caribs, but it is doubtful whether this innovation was not due to the fact that the Spanish conquerors offered sufficient inducements to prevent this warlike people eating their prisoners. When a lot of desirable trinkets could be obtained for a captive, it might easily follow that he would be reserved for sale.

The Carib, who is now almost extinct, was an example of a more powerful animal than the gentle Arawak. At enmity with the other tribes, he could not settle on a particular hunting ground, and therefore was a man-eater. It is doubtful, however, whether this was not the result of necessity, rather than choice, as he would rarely admit to his Dutch or English friends, that he had such a depraved appetite. The very fact that a little shame was felt shows that cannibalism was only exceptional and not, as was said by the Spaniards, so common as to be almost universal. For an animal to habitually prey upon his own species would be such a great drawback as to be virtually impossible in any harmonious natural arrangement, and we may therefore consider cannibalism as altogether exceptional even in the country which was called *Cannibalor Terra* by its discoverers.

To consider man as one of the species of living things in the forest, instead of the ruler over everything, may perhaps shock the sensibilities of the ultra conservative. In Europe and North America we see him carrying on such gigantic works that he seems to be a monarch indeed, but here, where nature is so much more rampant, it is the tree which comes to the front—everything else seems subordinate. Men are few, quadrupeds scarce, and it is only when we come to insects and plants that we find a dense population. These latter are everywhere, mutually utilising each other for their own purposes, and entirely ignoring the fact that there is such a personage as the "Lord of Creation." If it came to a dispute as to their respective powers, no doubt the forest giant would have the best of it. The Indian chops him down now and then, but only to make room for his offspring, that rise with tenfold energy. An ugly clearing has been made, which is unsightly for a time, but this is soon filled up and obliterated as if it had never been.

The Indian hardly cares to fight with his great rival. He chooses a place for his shelter on the sand-reef, where trees are few and easily kept down. Even when he makes a clearing for his wife's cassava field, it is kept open only for two or three years. When the forest commences its grand work of regaining this little portion of its domain, the red man finds the labour too arduous, and prefers to make another clearing. The richer the soil the more difficult it becomes to make any headway, and as cassava flourishes only on the most fertile spots, and soon exhausts these, the struggle would be almost useless, even if it were possible.

The habitation of the Indian is a very simple structure—only a shelter from the rain, raised on a few sticks. The largest houses are well-built, oblong sheds, thatched with palm leaves, under which are hung the hammocks, that serve for both chair and bed, being in fact the only real article of furniture. A shed, or benab as it is called, can be easily erected anywhere in the forest. When the Indians are travelling a temporary shelter is put up for the night in an hour or two, and even a good-sized dwelling can be erected in a day.

It follows, therefore, that Indian towns, or even villages, are practically unknown. A settlement may be kept up for twenty years or even longer, and again it may be abandoned at any time. For some reason or other, perhaps from an impulse which is more instinctive than anything else, the inhabitants leave one creek and go to another. All their movables can be carried in their canoes, or, if necessary, on the backs of the women. It almost seems as if their absolute necessaries are exactly fitted to their means of conveyance. A hammock, an earthen pot, a cassava plate, strainer and sifter, and one or two water bottles, make up a load which is by no means light, but which as nearly as possible reaches the point where only the "last straw" is wanting. Of course the man carries nothing except his weapons and the few ornaments he has on. His wife would despise him if he should demean himself by helping her in what belongs to her special province, as he would equally resent her bearing his weapons. If there is a baby, the woman carries it in a little hammock slung over one shoulder, so that it comes in front, and does not interfere with the load, which is borne on the back, and supported by a band across the forehead.

If it be true that "the boy is father to the

man" it must be of the greatest importance to the naturalist to know something of the Indian's childhood. His mother retires to the forest and brings him forth without assistance, returning home to her duties as if nothing particular had happened. What she feels it is impossible to say, but we may presume that the instinct of maternity is very strong, and overcomes everything else. Child-bearing is a matter of course to her, one of the peculiar duties of a woman, in which she neither requires help nor would accept it if offered. Her husband smiles in his quiet way when he sees the little one, and calmly prepares to do what he considers his duty. He must not hunt, shoot, or fell trees for some time, because there is an invisible connection between himself and the babe, whose spirit accompanies him in all his wanderings, and might be shot, chopped, or otherwise injured unwittingly. He therefore retires to his hammock, sometimes holding the little one, and receives the congratulations of his friends, as well as the advice of the elder members of the community. If he has occasion to travel he must not go very far, as the child spirit might get tired, and in passing a creek must first lay across it a little bridge, or bend a leaf into the shape of canoe for his companion. His wife looks after the cassava bread and pepper-pot, and assists the others in reminding her husband of his duties. No matter that they have to go without meat for a few days, the child's spirit must be preserved from harm. In this, as in many other things, the Indian wife gets on better than her more civilised sister. She knows her own duties and does them, never expecting any indulgence or assistance from her husband. He does not trespass on her domain, nor she on his, and consequently they rarely, if ever, have any disputes. The great drawback to all this, from one point of view, is the lack of sympathy. This feeling is one of those which is not only wanting in the Indian, but also in other uncultivated races. It is obviously the product of the higher evolution.

When the child is two or three days' old his mother goes to work in the cassava field, carrying him at her breast, in his little hammock. The little one hardly ever cries, nor does he give nearly so much trouble as the European baby. Even when sick, that cross and fretful disposition so conspicuous elsewhere is almost unknown. When dangerously ill the Piaiman is called in, who commences by charging the father with some neglect of duty as the cause. He had gone hunting too soon and shot the child's spirit, or wounded it in chopping down a tree. Something must be done, and the father must do it; he has unwittingly done the wrong, and will have to make reparation in some way. By direction of this half wizard, half physician, he therefore cuts several gashes on his breast, allows the blood to flow into a gourd, and, after mixing it with water, gives it to the babe as a medicine.

Here we probably get the first beginnings of what has formed such an important part of almost every religion—sacrifice. We injure others, and must suffer in return. Even races very high in the scale of civilisation cannot understand that when something is done it can never be recalled or obliterated. It is again the germ of that passion which leads to quarrels and fights, battles and wars. There is a feeling of satisfaction to the animal part of our nature when we resent or revenge an insult. "You hurt me and I'll hurt you," is the motto of the true savage, but the American Indian has gone a step farther. If he has unwittingly wronged his child he is ready to make every possible reparation, no matter what the cost is to himself.

The Indian baby soon begins to use his limbs. His mother lets him sprawl on the sand until he learns to creep, regardless of the fact that jiggers abound, and that they will penetrate the skin by hundreds, and produce unsightly ulcers and even dangerous sores. Poor little fellow! he looks miserable enough as he lies naked on the bare sand, and learns that his finger nails were made to scratch himself when what may be called the "sweet itching" is particularly troublesome. Occasionally, his mother makes a raid on the pests, picking them out one by one with a palm needle, or brass pin if she has one. But, notwithstanding these intermittent operations, and even as a result of the punctures, the child is more generally covered with sores and scars produced by sand and dirt. Sometimes these ulcers are alive with fly larvæ, and have to be dressed with hot ashes, which causes much pain, but even then the child does not scream.

Morning and evening he is taken down to the creek and encouraged to throw his limbs about until able to swim, which art he has often acquired before he can walk alone. If at all backward on his feet he is stimulated by the application of *hucu* ants. Taking one of these virulent creatures between his forefinger and thumb, his father lets it bite the little crawler until he strives with might and main to get away. The sight of one of the ants makes him get up and toddle away, soon after which he is running everywhere about the settlement, stealing the cassava bread or picking bits of meat from the barbecue. He is never beaten by his parents, and hardly interfered with in any way, even when he plays tricks on the dogs and other half-domesticated animals, which have free range of the settlement. No one cares if he injures them—he is to be a huntsman, and like most other boys has cruel instincts.

When a little older his father makes him a little wood-skin canoe and a pretty ornamented paddle, in which he goes wandering up and down the creek. He is also provided with a bow and arrows suited to his height and strength, with which he shoots at the dogs, monkeys, and parrots, and drives them to take shelter in the neighbouring bushes. Not being confined in any way, these tame animals can easily hide and return when he gets tired. Then he goes with his father and helps to lay the spring hooks on the creek, or assists in hunting deer and labba. He goes to no school, nor does he receive what is generally considered as an education. Yet, every day he is learning lessons which will be of the greatest importance to him in after life. There is only one profession to him, and he can never know too much about that. His father tells him but little, leaving the impression to be made by example rather than by precept.

Few children are to be seen at the Indian settlements, and therefore the boy has hardly any companions. The girls are women in everything but age and size, and follow their mothers; they take no part in the boys' games nor have they apparently any of their own. Naturally they learn the duties of the women as their brothers do those of the men. Like their parents, even the boys and girls take life seriously, and there is nothing like the romping so common among the children of more civilised races. Nevertheless "boys will be boys" even among the Guiana Indians, and have a few games. Most of these are more or less connected with their future duties, such as shooting and hunting. Among them is one which can be played by four or more, and is a life-like imitation of a labba hunt. One boy represents the labba, a second the dog, and the third and fourth two huntsmen, one of which takes up his station in a little canoe on the creek, while the other takes the so-called dog and enters the forest. The "labba" is given a little time to hide, and then the hunt begins, the "dog" barking and the huntsman calling to the other in the canoe as the quarry doubles in and out among the bushes. Presently the supposed animal makes for the creek, plunges in and dives under water, followed by the "dog" and huntsman, who with his assistant in the canoe, tries to lay hold of the diver. For some minutes the water is all in a commotion as the three naked urchins swim about, the "labba" doubling in every direction, now under water in one place, and suddenly rising to the surface in another, the boy in the canoe paddling with all his might until the supposed game is taken and the play ended.

Ball playing is carried on by men rather than boys, but there is a game of shuttlecock sometimes engaged in by the youngsters. Corn cobs are stripped, and three feathers stuck into one end. The object of each player is to throw his shuttlecock as high as possible, so that it may gyrate the longest. In a village some five hundred miles from the coast children have been seen shooting at marks with home-made cross-bows, but this weapon has obviously been adopted from the early voyagers, as it appears to have been unknown to the aborigines.

Whatever semblance of teaching the boy receives from his father is little more than the mode of handling a gun, bow, or blow-pipe, and even that is learnt more by imitation and practice than by precept. The only attempt he makes at writing or drawing is to scratch figures on the sand with a piece of stick like those of his mother's bead apron, or conventional outlines of animals such as may be seen on the two or three benches which are used to squat upon and keep the feet raised above the sand and its multitude of jiggers. He learns the names of the game animals, birds and useful plants, by hearing his father talk of them, and their uses by practical experience. From that tree his bark canoe was peeled, from this kind his father's dug-out was made; here is the bow-wood, and there that from which he got his paddles. When walking through the forest he sees his father bend down a twig here and there as a guide on his return, and the boy does the same. Then his eyes begin to appreciate the various things which go to make up a knowledge of bush travelling, such as the lay of the land, the watersheds of particular creeks, and certain prominent trees. Like great public buildings some immense forest trees stand out prominently above the others and become landmarks, and others get known from certain events which happened in connection with them. Among the roots of this great mora his father caught an acourie, and in that hollow log a labba. The blasted trees half-strangled by wild figs are favourite resorts for parrots and toucans, and underneath the saouari-nut tree he may look for the footprints of many game animals. By and by he finds out the seasons when nuts and fruit ripen, and in connection with these when he ought to go hunting or fishing in particular localities. Like other boys he has a liking for many wild fruits and nuts, and knows where to look for the fat caterpillars which infest the leaf-buds of so many palms. These he eats with a great relish, as he does also certain chrysalids found in the cassava field, cracking the latter as an English boy would hazel nuts.

Now he is given a name, which hitherto he had not possessed, nor even yet does it appear that there is any particular use for it as it is carefully suppressed. He will still be called "boy," and later "friend" or "brother," never by his proper appellation, which is a secret to all but his immediate relations. It is so difficult to get at these names that travellers who inquire are generally put off with some general term of relationship such as father, wife, or son. The name is so sacred that the Indian becomes sullen when you insist upon learning it, and the only time we ever saw one of them in a passion was after our asking a number of questions in regard to the matter and refusing to be put off with evasions. It is generally supposed that the name is considered so intimately connected with the personality that to know it is as bad as our ancestors thought it to let a witch have a lock of their hair. If generally known perhaps some enemy might get hold of it and do the owner much harm. Whether there is any other superstition connected with the name is doubtful as the whole subject is so very obscure. It appears that boys are given the names of game animals, and girls those of trees and birds, but we make this statement with some hesitation, as on this matter we have certainly been purposely deceived on more than one occasion. Again, it might be suspected, that something like the totem system of the North Americans exists here, but this is also obscure. We have heard of cases where an Indian has been debarred from killing a certain animal at particular times, as, for example, when a child is born, and suspect that this is connected with his name; but we can do no more than put it as a problem to be solved, if possible, in the future. Writers on these people generally assume that they have reasons for everything they do, but this is hardly ever the case. Like children, they do many things from impulse, and are governed by very powerful instincts for which neither they themselves nor any one else can find reason or excuse. The fallacy, *post hoc, ergo propter hoc*, is engrained upon their minds, and if we could possibly get at the origin of most of their customs, we should no doubt find that they originated in something easily explainable. In the absence of such evidence, however, we are entirely in the dark, and it becomes useless to attempt anything like explanations of the why and wherefore of these things. Uncivilised races everywhere have manners and customs for which they never attempt to give reasons. Perhaps some plausible explanation may be given to a stranger, but in many cases it has never occurred to the people themselves that a reason was necessary.

Very little gossip is carried on at the Indian settlements. Outside their own little community they have no interests whatever, and therefore all the conversation is connected with their personal experiences. Women talk less than men, as they have no hunting or fishing exploits to recount. This reminds us that the Indian female differs in many other respects from her civilised sister than in holding her tongue. She hardly ever wears ornaments—these belong to her husband. He paints and decorates himself from head to foot, while his wife has only her pretty bead apron.

His feather crown and belt glow with colour and show very good taste, and even his paint is not glaringly discordant. Necklaces of peccaries' teeth and seeds, with tufts of toucans' breasts hanging down the back; bracelets and anklets of beetles' wings; and belts of rattling seeds, all go to decorate the man, leaving the woman without even a floral wreath. Sexual selection appears here as in most of the lower animals to come from the female side; the male decorates himself, dances, boasts of his prowess in hunting, and plays Othello as a matter of course, while the female tacitly approves and gives him her homage.

Life is a serious thing to the Indian boy. He is ambitious to become a man as soon as possible. Around him he sees skilful hunters and fishermen, and wants to be like them. His father tells him to wait—he is not yet a man, nor can he become one until he has gone through the proper ordeal. Of course, he wants to do this as soon as possible, and is proud to show how well he can shoot, paddle, and chop down trees. Then, to prove his capability of bearing pain he allows manourie ants to bite him, or cuts gashes on his arms and breast, into which he rubs the acrid juice of a beena in imitation of his father and the other men.

This reminds us that but little is known of this

peculiar custom of inoculating with the juices of acrid plants. The Creoles of Surinam have apparently adopted from the Indians, through the bush negroes, a supposed protection against snake bites, which they confidently believe not only to prevent these venomous creatures from biting them, but also to ward off any evil effects should they do so. The principal ingredient used is the pounded head of a labaria, or rattlesnake, mixed with the juices of certain plants, and this is generally rubbed into an incision on the wrist or fore-arm. No doubt the confidence resulting from perfect faith in this remedy enables the Creole to handle the snake with impunity, while cases of recovery from bites, which are by no means rare, add to its reputation. What the Indian thinks of his beenas, however, it is almost impossible to say; possibly he never thinks at all, but simply uses them because his ancestors have done so for ages. In a general way he believes that the fact of his having been inoculated, for example, with the jaguar beena, promotes his success in hunting that animal, but how it can do so is another matter. It has been suggested that he thinks it makes him invisible, less conspicuous, or perhaps covers his scent, thus allowing him to approach within a very short distance of his game without alarming it. As this is a very important matter to the Indian, who is by no means a good shot at a long range, he would naturally attach great importance to anything that would help him towards this end.

Near the settlement are always planted a few beenas. A stranger, when he sees a clump of the scarlet amaryllis (*Hippeastrum equestre*), or a pretty collection of caladiums with different markings on their arrow-shaped leaves, would think the wild man of the forest had some sense of beauty in form and colour. He is certainly not wanting in taste, as may be seen from the arrangement of colours in his feather crown, but even if this had anything to do with the selection of these beautiful plants in some bygone age, it does not appear to exist to-day. An Indian wears no floral garlands, nor does he seem to appreciate plants at all, otherwise than for their utility. However, they are there—flowers enlivening the bare sand with a fiery glow, and leaves mottled with white, pink, or crimson—ready for use if the hunter fails in shooting a tapir or labba. The poor fellow comes home much depressed—his virtue has gone out of him. Years before he had been inoculated for the jaguar, deer, all the cavies, and even the alligator—perhaps the charm has worn out. Going to his clump of beenas he takes up a bulb or tuber, and after making several gashes over the breast and arms, rubs in the acrid juice without hesitation. Although the operation is very painful he does not even wince, but seems to glory in his stoicism. Next day he has regained his confidence, goes forth into the forest, is successful, and of course puts everything to the credit of the beena.

In addition to these plants there is another beena which has to be used before our boy can arrive at manhood. This seems to be a general charm to make him successful in hunting and fishing, as well as in all undertakings worthy of the future head of a family. Here, again, we are met by the difficulty of accounting for the use of such an instrument. Of the size and length of a coach whip without its handle, the nose beena looks as if intended as an instrument of punishment. And so it is, in a way, but not as might be at first sight supposed. It is well greased with the fat of palm caterpillars, the thin end pushed up one of the nostrils into the air passage, and drawn out through the mouth. The boy endures the pain and irritation without flinching, and after giving this and other proofs of his endurance is no longer a child.

There is one other manly accomplishment which the young Indian has to learn, and that is to drink piwaree—a fermented liquor made from chewed cassava bread. For want of other vessels this drink is generally made in a "dug-out" canoe, and when the time comes for the drinking bout the quantity which a few men can dispose of is astonishing. Calabashes full are handed round by the women, and the men appear to emulate each other in the number they take. Gallon after gallon disappears down their throats, and every now and again one goes outside to eject what he has just taken. Altogether, these orgies are a blot on the otherwise admirable character of the red man, and show up his one weak point—a taste for drink and an intense craving for the pleasures of intoxication.

However, we will not dwell upon this unpleasant subject, but come to the climax of the youth's ambition. He cannot be altogether a man until he has a wife. With a feather crown on his head, a necklace of peccaries' teeth round his neck, and perhaps a feather belt below his waist, he struts about glorying in the fact that he is no longer a child. The young girls look on him with admiration; they have seen his bearing under the ordeal. Very little courting is necessary: the couple agree, and, as a matter of course, the bride hangs her hammock beside that of her husband the same evening. There may be some arrangement between the parents, but very often this is dispensed with altogether, and the union is hardly noticed.

Having followed the Indian child to manhood, we can now leave him. He has attained the position for which he has been striving, and can now go on with the two duties which every living thing has to perform, first, to struggle for food, and, second, to see that his line does not become extinct. During the last century his old environment has been somewhat changed, so that progressive development is almost arrested. A great factor in evolution, war, has come to an end, and although he still has to hunt and fish, he no longer fights with his fellow man. This condition, which, at first sight, might be looked upon with complacency, is evidently unnatural, and its results are shown in deterioration and gradual extermination of the race. The Indian cannot hold his own in presence of the white man, but retires into the forest, where, having no incentive to do more than work for his daily food, he dies out. Man, like other animals and even plants, has to fight in some way or other, it may be with the elements or it may be with his fellows, and he who has had the most difficulties is the strongest and fittest to survive. If happiness, as we said before, consists in perfect accommodation to the environment, then the Indian is the happiest of men; but this naturally brings up the question whether it is good for man to be happy. Nature answers most emphatically, No; but at the same time tells us to strive for its attainment. If it were possible to conceive of the attainment of perfection, with nothing left to hope for, we could only think of such a state as the extreme of dulness. Even if we lost every desire and aspiration and could settle down to a state of do-nothing for our life-times, it would still be a particularly unenviable condition. Even rest is the complement of labour, the one is only possible in connection with the other.

It seems pitiful that the Guiana Indian should be exterminated, but, nevertheless, this end is certain. Unless the whole country were abandoned by the European he could never again come to the front, and even if such an unlikely thing ever came to pass, the consequences of present interference would probably affect the result. During the last few years the gold prospector has been intruding within the Indian's domains, with the result that he retires farther and farther away, often carrying with him the germs of diseases unknown to his ancestors. Formerly, he had remedies for all sicknesses with which he was acquainted, and often showed considerable knowledge of the virtues of certain medicines—now small-pox, syphilis, and the evil effects of rum have put all this out. It has taken him ages to learn what remedies are most suitable for his peculiar diseases, and as he cannot learn from books, and refuses to take advice, there is absolutely nothing to be done. Where once the Indian families were enumerated by thousands there is now not a single individual, and a few years make great differences even in the far interior. Villages where once the traveller found a hearty welcome can hardly be distinguished from the surrounding forest, not even a parrot remaining to speak the language of the lost tribe.

III.

THE ANIMALS OF THE FOREST.

To the stranger the forest appears almost deserted. Hardly the sign of an animal is to be seen by any but a skilled huntsman, and by him only after a most careful search. There are no open places, but the whole is one vast game cover, in the recesses of which millions of animals may be hidden without an indication of their presence. For there are no herbivorous animals, as there are no pastures on which they can graze. Not a blade of grass or hardly a green leaf is seen under the wide-stretching roof, and it would therefore be impossible for them to live. Even the deer of the savannah, although sometimes found in the jungle, cannot exist inside the forest. It has followed, therefore, that in the course of ages the wild beasts have accommodated themselves perfectly to their environment, and are now as well fitted to it as the trees themselves. Nowhere, perhaps, in the whole world are there so many distinct forms of animal life as in South America. Here is the home of the cavies, that family of which the well-known but miscalled guinea-pig is one of the smallest members. Living on the nuts which strew the ground in such profusion, these pretty creatures abound in the forest. Hidden away in hollow logs or among the great tangle of roots during the day, they come out at night to nibble at the monkey-nuts and other seeds. Then there is the tapir and two species of peccary, which, like their cousins, the hogs, live almost entirely on the seeds that are scattered so profusely below by the great forest giants. These animals are also nocturnal in their habits, generally hiding away in some dense tangle on the sand-reef during the day. To prevent their settling down too easily, nature has provided enemies in the shape of that beautiful series of wild cats which culminates in the jaguar or American tiger. Like their cousins, so well known in civilised countries, they are good climbers, can see well at night, and often spring upon the luckless acouries or labbas as they are quietly feeding. Now and again one of them catches a peccary which has strayed from the herd, but the wild hog has learnt that there is safety in combination and rarely strays from his fellows.

Among the most curious of the inhabitants of the forest that are not arboreal is the ant bear. This creature, with its powerful forelegs, great claws, and tapering snout, suggests at once that there have been several strange factors at work to create such a monster. And, when we find that it lives entirely on ants and termites, our wonder is by no means diminished. To feed an animal of this size on such pigmies seems almost unnatural, and yet the multitude of ants in the forest and on the sand-reef is so enormous, that there is, after all, no difficulty in realising the fact. We shall have something to say of the ant world presently, so will now only call attention to the impossibility of studying one part of nature without some knowledge of other divisions.

In the silence of the forest the least sound is heard by the animals long before a noisy intruder can get a sight of them. The labba peeps from his home in the hollow tree, cocks up his ears, and, if he hears a footstep, retires again to its recesses. If near to the water he dives and comes up some distance away, hiding his head among water-plants or in the midst of one of the dense bushes which come down into the stream. The acourie or agouti, however, does not dive, but, like the fox, is very wily. Going in at one end of a log and out at the other, he often escapes while the huntsman is looking round. If chased he will run along the shallows of a creek to hide his scent from the dogs, or swim over and back again several times for the same purpose. He never runs straight when pursued, but doubles, often hiding until a dog has passed and then making off in a different direction. Like the fox he has been hunted for a very long period, and like Reynard has grown wiser with every generation. The accumulated experience of past ages has made him as cunning to evade his pursuer as it has made the Indian and his dog knowing in their trade as hunters.

Most of the forest animals living on the ground readily take to the water, in fact, some are almost amphibious. The deer, however, although he swims easily, cannot dive or move fast in that element. His home is on the savannah, and when there he can easily get away from the dog, but if driven to cross a broad river he is lost. The Indians hunt him as they do some other game, one man with his dog driving him to the creek, where another stands ready in a canoe to capture the poor creature as he goes to cross the stream.

His lair is among the bushes on the edge of the forest, and like most other animals his colour harmonises so closely with his surroundings, that he is practically invisible to any one but the Indian.

This invisibility is a striking characteristic of every living thing in the forest. At first a stranger observes nothing but a scene of desolate confusion. Later, however, he begins to distinguish one tree from another and learns where to look for a particular animal. Then he wonders how he could have missed the signs which now impress themselves upon his eyes. However, this is not altogether characteristic of the forest as a similar result follows on a close acquaintance with any place. There is another aspect of this question, and that is the fact that there must be a desire for knowledge. Otherwise, the forest is excessively tame and dull—perhaps to an ordinary sojourner almost as lifeless as the African desert. For want of knowledge he sees nothing, hears nothing, and is inclined to do nothing but complain of the monotony day after day. There is nothing to relieve him from the feeling of annoyance produced by the hot, steamy atmosphere and the insect pests which continually worry him. He feels clammy; his clothes get damp and cannot be dried; in one place there are mosquitoes and in another sand-flies. And then the ants. They are everywhere. They crawl about under his clothes, nip him about the neck and arms and ankles, get into his food, perhaps carry away his sugar grain by grain in a night, and altogether are most annoying. All these things, however, help to make the bush most fascinating to a naturalist; he is no longer a looker-on at the show, but an actor in the midst of it. At every fresh visit he finds something new—something to think of when he returns home—some problem to be inquired into on another excursion. No matter that he rarely succeeds in solving it—the eager desire and hope keep up the interest.

But we are wandering off into that maze which is so interminable, and must come back to the animals of the forest. However rare and difficult to find may be those which live on the ground, still they are to be seen by the naturalist and skilled huntsman; but when we come to the others —the great majority that abide in the canopy above and rarely descend to earth—observation is almost impossible. Had we the wings of a bird we might hover above the tree tops and see their inhabitants living and enjoying life, as beautifully fitted to their environment as the others; without such appendages we must be content to glean a little information now and then. If the cover for ground game is so very great, what shall we say of that for the monkeys, opossums, sloths, iguanas, snakes, and birds. From the din some of these make in the early morning we might suppose them to be the only inhabitants of the forest. And how weird are their cries! They add to the feeling of awe which is almost inseparable from the dense shades. The red howling monkey, hidden in the foliage overhead, keeps up his reverberating notes at intervals for hours, and makes the stranger exclaim almost in a fright, "Whatever can that be?" Then come the tree-frogs, which astonish us with their loud whistling or booming, while the buzzing of the cicada or razor-grinder is even more startling. Near the settlements the latter is called the "six-o'clock bee," from its characteristic noise being heard at that hour, as it flies from tree to tree. Suddenly, without warning, you hear a grindstone, as it were, at work in the tree overhead, and presently, if the canopy is not too thick, see a great fly pass quickly into another tree and repeat its peculiar buzz. In the forest at certain seasons these insects are heard all day, but instead of enlivening the awful stillness, they rather add to its solemnity. Those who have wandered alone over some mountain slope and started as a sheep hidden behind a boulder uttered its prolonged baa, will understand how these strange noises affect us in the forest. But however startling such a sound may be on the mountain, it is much more so here, for not only is there silence, but gloom, and the reverberations are increased by echoes from the myriad treetrunks.

At night a continual hum is heard, which however does not enliven the forest. It is like a singing in the ears, rather pleasant than otherwise, and is produced by myriads of insects which fly around after sunset. If you are paddling up a broad river and keep in the middle of the stream, the hum is imperceptible; but on approaching the shore it becomes almost piercing until you become accustomed to it and then it is hardly noticed. Now and again an owl or goatsucker flits past, uttering its weird cry, which is even more startling than that of the red howler. Suddenly you are startled by the question, ' Who are you?" or told to " Whip poor Will!" Then comes that series of wails which Waterton compared to the midnight cry of some murdered victim, waking the sleeper and sending a cold thrill down his back.

These cries are supplemented or replaced in the early morning by a din above our heads, which at once proves that the forest is by no means wanting in animal life. From the top of a giant mora comes the screeching of a flock of parrots or macaws, here and there a toucan is uttering its puppy-like bark, from far away comes the ting of the campanero, and a hundred other species utter their peculiar cries at intervals. Before sunrise they begin to wake; here and there a bird greets the morn like the domestic cock, and as the sun begins to flash his rays over the green expanse, every bird is up and doing, his brilliant colours glowing against the almost black foliage. But even then a sportsman complains that they are few. In an English wood he looks for and generally finds flocks of one species of bird; but here in South America, with few exceptions, he has to be content with individuals or very small companies.

When we see the average white sportsman go into the forest to destroy such beautiful creatures as parrots and toucans, simply to gratify his murderous appetite we are more than shocked. Even when he wounds one of them it often clutches at a twig and remains there to suffer, or drops in the forest where it cannot be found. The Indian kills to eat, and very rarely wounds. He cannot afford to waste powder and shot, while the amateur sportsman is careless of this. The consequence is an amount of suffering which he surely cannot appreciate, or he would be afraid to lift his gun against a parrot. The naturalist who has such a man for a companion on a bush excursion is horrified to see him let fly at anything and everything—not even those little wanderers from paradise, the humming-birds, being excepted.

Snakes are plentiful on the edge of the forest, but they are rarely seen by any one but the naturalist. Although often so brilliant in the light, otherwise their colours assimilate to the bark of the trees round which they coil. As a rule they are very sluggish, although able to move very swiftly when frightened. By walking quietly like the Indian, however, we may often see them with their heads peeping out from among the branches, as if on the look-out, sunning themselves across some forest path, or taking a drink from the creek. As harmless to man as they are beautiful in his eyes, when he gets over his natural repugnance to them, the naturalist often wonders why snakes are held in such abhorrence. True, the venomous species will turn when trod upon, but they are not to blame for that. Without their poisonous fangs the species which live in the trees might often find it difficult to get a meal.

Those living on the ground have the advantage of being able to dart on their prey and coil round it so quickly that the movement is hardly distinguishable. With their bodies coiled round a branch this is not so easy to the tree-snake, and we may therefore presume that the poison fangs were originally developed to get over this difficulty. It is true that there are many non-poisonous species now living in trees, and a number of venomous kinds at home on the ground, but it can be easily conceived that the first development may have taken place in an arboreal species. At first sight the naturalist may be inclined to pity these poor creatures, as they seem to have so few weapons of defence, but when he studies them a little more he is struck with their many beautiful adaptations to circumstances. Without proper teeth and entirely devoid of claws they seem almost helpless, yet they exist, develop, and show no signs of degradation, but, on the contrary, appear to flourish to perfection.

Lizards and frogs, like snakes, appear to have been originally confined to the ground and water, but in Guiana some species that live in the trees are among the most interesting of our animals. As in the case of other forms of life these show marvellous adaptations to their environment, and should always be studied in this connection. Guiana is pre-eminently a land of forest and stream, and it has followed that both animal and vegetable kingdoms have been developed to suit these conditions. Some are equally at home on land, in the water, or on the trees, those that cannot easily live in the flood being able to climb out of its reach. Then we must also take into account the kinds of food procurable. The interdependence of one animal on another, and these again upon the seeds of trees and even on flowers, is so close, that we can hardly conceive of their existing apart. Changes of environment in past ages have undoubtedly caused the extinction of numberless species, and from the naturalist's point of view it would no doubt be a great catastrophe were the South American forest even partially cleared.

If the larger animals are wonderfully adapted to their habitats, the insects are particularly conspicuous in the same way. We have seen that game is, at least apparently, scarce, but we cannot say the same of the smaller forms of life. Like the plants, they are not only varied in species, but exceedingly numerous from every point of view. Here we find the most magnificent beetles, moths, butterflies, and ephemeræ, exceedingly curious forms of mantidæ, grasshoppers, flies, and spiders, and an almost bewildering variety of wasps and ants. In the day the butterflies, ephemeræ, wasps and bees, are as conspicuous as the flowers, while the ants are really everywhere. On the sand-reefs and high ground generally, the tracks of those species of ants which live in the earth are seen crossing and intersecting each other in every direction. Processions are continually passing and repassing, those that gather leaves to make their mushroom beds reminding us of some great Irish national *fête*, where every one carries a green banner. Then comes the great army of hunters or scavengers which frighten and scatter every living thing on their route. They rummage out every chink and cranny, now catching a fat beetle-larva, now a cockroach, and farther on, perhaps, some sick or wounded animal. In vain the cockroach tries to run or even fly; several of the ants have taken hold, and will let themselves be torn in pieces rather than lose their advantage. Almost before you can see what is going on, every particle of the soft flesh of the insect is eaten, and nothing remains but the two wing-sheaths and the covering of the thorax.

Here we have in the same class species that are carnivorous as well as others distinctly herbivorous.

The latter live in well-built nests in the ground, as well protected against floods and enemies as if the ants formed great nations with lines of dams and fortresses. But there are other species found on the trees, all having their peculiar manners and customs, which should be almost as interesting to the student as those of the races of mankind. A small black ant which lives in the Barbados cherry (*Malpighia punicifolia*) builds itself a little nest the size of a walnut, which it fastens to a twig. From this home it wanders over the bush, apparently looking after flocks of scale insects which it has carried to pasture on the leaves when very young or in the egg. The scale insect sucks the juices from the leaf, and the ants crowd round seemingly to get a portion of what exudes, or else to take it from the parasite after some change has been effected.

Some trees provide homes for these little creatures, evidently inviting them to inhabit their barracks, provided they keep off noxious creatures that would eat their leaves or flowers. A species of Melastomaceæ does this in the swollen petioles of its leaves, an acacia in its thorns, another shrub in a swollen node of its branches, and an orchid (*Diacrium bicornutum*) in its psuedo-bulbs. All these are so beautifully contrived that we can hardly think of them in any other light than as provided specially for the purpose. Then there are less elaborate contrivances, the most perfect of which is that of the Coryanthes, which, unlike many other orchids, lets its aerial roots grow into an oval ball where ants can easily take up their abode and fill in the lattice-like spaces to make a perfectly safe habitation and barracks. Other epiphytes provide more or less perfect mats of roots in which ants also make their homes, and sally out in defence if the plant is disturbed. Heaps of fallen leaves, matted stems of creepers, or any collection of *débris* at the forking of tree branches are also utilised by these interesting creatures. Whether they ever sleep is doubtful—they crawl into our hammocks at night, and drop upon us as we brush past the hanging bush-ropes by day. Some, as we have seen, combine for a special object; others appear to work independently. Hardly a trunk, branch, leaf, or flower, is free from them. They vary in colour from bright red, through brown to jet black, and in size from little creatures that are almost microscopic to monsters nearly as large as wasps. Like snakes, many of them instil a poison when they bite, while others nip tiny pieces of flesh from their victims, and are not venomous. The manuri, which is perhaps the largest species in Guiana, is over an inch long and is so venomous that the Indian uses it to test his boy's capacity of bearing pain. We have heard a negro cry out and almost weep from its bite, the effects of which lasted for hours. To study these interesting creatures would be the work of a life-time. The researches of Sir John Lubbock have thrown a flood of light upon the habits of European species, but the field here is so wide that probably the result of similar researches in Guiana would be marvellous.

Perhaps, after the ants, the most ubiquitous are the termites, who are the scavengers of the forest. Although, perhaps, among the most helpless of living things, they carry on a work which is of the utmost importance. In a year or two they will break down the largest timber tree, until it collapses into that rich brown humus so characteristic of the forest, and which is so well suited to feed all the other plants. It must be understood that the decomposition which takes place in the forest is quite distinct from that of more open places. Instead of the alternations of rain and sunshine, we have here a uniform temperature and almost equal amount of moisture. It follows, therefore, that the decomposition is even and continuous. The termites burrow through the timber in every direction, allow the moisture to penetrate, and in a comparatively short time the hard wood, which rings almost like metal to the blows of the axe, crunches under foot as if it were made up of eggshells. As long as a tree is healthy the termite leaves it alone, but as soon as a branch is injured a nest is sure to be planted in the fork and its work begins. As there are always plenty of the dead and dying in the forest, these little creatures may be seen everywhere, plainly indicating by their presence that another poor victim has succumbed in the struggle.

Wasps and bees are also very numerous, the former hanging their round or pear-shaped nests from the branches of trees, or even building a single layer of cells on the backs of leaves. They are called marabuntas in the colony, and are much feared for their virulent stings. Unlike the wasps of Europe they do not appear to live upon fruit, but to be carnivorous, or general scavengers, like the ants. Although not so interesting as the bees and ants, they are well worth study, and no doubt careful investigation would be amply rewarded. Bees are not by any means so common as the wasps, although the humble bees are more plentiful, as might be expected from their being so well fitted for the fertilisation of flowers.

Coming now to the nocturnal insects, which include flies, beetles, and moths, we are almost bewildered by their number. As we have before mentioned, their buzzing is kept up all night long, even where there are no mosquitoes, which undoubtedly help to make up the din when they are present. Hang a lantern or candle in your forest camp, and a cloud of insects are attracted by the light. First come the gnats and other flies, which fall into an open flame and create quite a litter underneath. Then the smaller moths burn their wings and drop, while every now and again a great sphinx comes fluttering along, or a monster beetle flies straight and strikes against anything in its way.

With so many insects which pass through their first stages in the water, it naturally follows that the swamp, river, and creek, are teeming with life. The lower animalculæ flourish everywhere and at all times, to feed the myriads of mosquito and fly larvæ; these again nourish the shoals of smaller fish, which in their turn fall a prey to their large finny brethren, as well as to the alligator and the ibis.

IV.

INTERDEPENDENCE OF PLANTS AND ANIMALS.

ONE day in passing along a creek, we had an experience which set us a-thinking. We were collecting orchids, and up in a tree overhanging the water was lodged a great clump of Oncidium altissimum, its long graceful flower-stems loaded with yellow butterfly-like blossoms hanging over in every direction. It was a magnificent plant, fully four feet thick, with panicles rising to a height of twelve feet. It is needless to say that we wanted it for our collection, and that we sent one of our boatmen to fetch it down as carefully as possible. This, however, was easier said than done, for, first it was attached to the tree, then it was threaded, as it were, with a number of bush-ropes, and finally more or less entangled in a crowd of branches. Taking a cutlass, the negro climbed up to its level and began to chop at the obstruction, but almost immediately came down with a run, rubbing his hands and face and picking a swarm of ants from his clothes. Looking up we could see that the attack on the plant had brought out its garrison, which blackened every leaf and flower-stem, and made the negro descend in such hot haste. We will not go into details as to how we procured the plant at last, but only mention that the ants kept us at bay for fully half an hour before we could throw it into the stream. Then came the work of getting rid of the virulent insects by pushing the whole mass under water with a long bamboo, and keeping our bateau upstream to prevent their coming on board. As the roots became soaked the creek became covered with black patches, and it was quite a work to keep them from running along the bamboo. Presently, however, several larger forms were seen swimming in the water, and these turned out to be cockroaches, which also went floating down with the ants. But we had not yet dislodged all the occupants, for soon a large centipede was seen struggling in the crowd, and, as may be supposed, we did not hurry to take in the plant until sure that this was the only one of its kind harboured by the Oncidium.

One of the party suggested that this was "a

happy family," but we could hardly agree with him if he meant that they dwelt at peace with each other. To us it brought up the great problem of the dependence of plants on the natural elements, of animals on plants, and these again on other animals. By means of the soil, water, air, and light, the forest tree rose beside the creek, and up it had grown the elegant bignonia, whose flowers were hanging so gracefully far overhead, and whose stems stretched like cords from base to summit. On a fork of the tree, and among the bush-ropes, the orchid had found a congenial habitat, where it grew and flourished for years, developing a great mass of roots to be occupied by the immense horde of ants. The plant might perhaps have lived without such tenants, but it is most probable that its flourishing condition was mainly due to these little creatures. For, they were there with consent, and in return for house accommodation undertook to keep off the enemies of the orchid, of which the cockroach was one of the most inveterate. Why, then, were these pests allowed to be present? We can only suppose that the attraction of the plant drew them, and that they had not yet been captured. For it is not to be supposed that even this omnivorous insect will be deterred from attempting to get a delicious meal simply because its enemies are in the fore. Again, was not that one of the reasons why the ants were so ready to take up their abode among the orchid roots? Where its food was to be found the cockroach would certainly come, and the ant as certainly find its prey. And what shall we say of the centipede? Like the ant it loves a fat cockroach, and was present in hopes of finding one.

We might carry this example of interdependence still farther, although the other developments did not come so immediately under our notice. The tree was an etabally (*Vochysia*), and far above us exhibited a glow of yellow from its being literally covered with golden blossoms, over which were flitting hundreds of sulphur-coloured butterflies, hard at work sipping nectar, and at the same time carrying on the grand work of fertilisation. The tree is a conspicuous object from a long distance, the butterflies are attracted to it in hopes of procuring food, in sipping the nectar they fertilise the flower and thus enable the etaballby to procreate its species. Then it might easily happen, although we cannot say that it did in this particular case, that the first stage in the life of the insect was passed on the same tree. The butterfly sips the nectar and then deposits its eggs on the under surface of the leaves, from whence come a host of larvæ to spread devastation all around, perhaps leaving the tree almost bare. But this is little more than a rough pruning, which causes it to flower all the more freely and produce a greater supply of nectar for the perfect insect.

Advancing another step we see that the seed produced through the medium of the butterfly attracts birds and monkeys in the day, with bats above and rodents beneath at night, those above playing havoc among the branches, quarrelling and fighting with each other for the fruit and dropping them by thousands to feed the nocturnal prowlers. Finding the tree so convenient the birds pair, build their nests and bring up their young upon it. Now for another aspect: the host of caterpillars bring a crowd of insectivorous birds, which also make themselves at home in the midst of such a grand feast. These are accompanied by a host of flies which have scented their prey from afar, and now come to help carry on the work of preventing the larvæ from going too far; and, because the flies are there, the goatsucker comes hovering round at night. Unlike the birds, however, the flies do not kill outright, but, as it were, take possession of them for the benefit of their offspring. Piercing the skin of a larva they deposit their eggs underneath, so that when they hatch the young have free range of their host, and eat up everything but what is then the membraneous covering of a pupa.

Now we come to a further aspect of this interdependence. Hawks soar overhead, and hearing the loud chatter of parrots and other birds, now and again pounce down and drive them to hide in the densest part of the canopy of foliage. But, even here they are met by a new misfortune, for snakes have climbed from below to get their share of the good things, and are ready to pounce upon them as they flutter away from the hawk. Down below, in the lower branches or on the ground, the jaguar, puma, or ocelot, lies in wait, and he also lives because that cloud of yellow butterflies gambolled about the flowers a few months ago.

Besides all these the seeds and fruit of forest trees go to feed the shoals of fish which make their way from the great rivers, and swim about everywhere during the flood. On account of their presence the great jabiru, or giant stork, frequents the inundated tracts, and vies with the alligator and the Indian for a share of the finny spoil.

We might go on further and tell of the weevils which bore into the fruit as it lies on the ground, of the cockroaches and great beetle larvæ with their parasites, and of the ants and scorpions which hunt these, and of the thousand animaculæ which also obtain their share of what nature has lavished so freely. All these get their fill, and yet there are always more seeds left than can find room to grow. But we will go on to another scene.

In the savannah rises a great eta palm, perhaps sixty feet high, its mass of roots standing above the water as a mound, from whence proceed the rough but bare stem to a height of fifty feet, where the great dome of fan-shaped leaves crowns its apex. Here are no rivals of its own kind, no bush ropes or smothering creepers, and hardly anything to dispute its claim as monarch of all it surveys. Even here, however, are signs of interdependence. Below the crown stand the remains of a hundred clasping leaf-stalks of different ages, their axils filled with decaying vegetable matter, in which revel the aerial roots of that unique orchid, Catasetum longifolium. With ribbon-like, flexible leaves streaming downwards and great flower-spikes slightly bent outwards to greet the sunlight, this plant also appears to have no rivals. However, we want the orchids and must get them, and the only way to do this safely is by cutting down the beautiful palm. We regret this necessity, and even go so far as to send one of the negroes (a well-known cocoanut gatherer) to bring down a plant. But he gets startled at a small gecko lizard, and with a

cry of "A wood-slave bite me," comes down with
nothing but one of the Catasetum bulbs, which
he has hurriedly torn off in his fright. The poor
little reptile could not injure him in any way, but
as our specimens would be useless if torn to pieces,
we resolve to cut down the palm. At first the
axe rings on the hard trunk, as if both instead of
one were made of steel, but presently, as an entrance
is made, the wood proves quite soft. Then the
mighty prince of the vegetable kingdom, as Lin-
næus would have called it, bends over, comes crash-
ing down, and throws up sheets of water and mud
as it strikes the surface of the swamp. We wade
towards the crown and begin feeling below for the
Catasetums, and presently notice that the water is
black with ants, which soon make their presence
known most unmistakably by their virulent bites.
However, we are not to be daunted by these pests,
and soon manage to loosen an orchid, bringing
up with it a nest of hairy spiders, the dreaded
wood-slave, a little harmless snake, quite a number
of cockroaches, and two or three beetles. We also
discover that several plants have entered into competition with the Catasetums, including a small
species of Vanilla, a few ferns, and one or two
Gesneriæ.

Like an island in the sea this palm crown stood

far away from dry land, till it became a little world in itself, with carnivorous and herbivorous animals, and plants, all living, fighting, and killing each other, but still keeping up the balance of life.

Although we are continually speaking of "the struggle for existence" and "survival of the fittest," few, it seems, are able to appreciate what these sentences mean, but to the naturalist in the forest they are full of suggestions. If it were possible to have absolute peace throughout a world where there was no difficulty, sickness, or death, it could only consist with a dulness of which we can hardly have any conception. From the lowest plant to the highest animal, all have to work hard and get material to build up and keep alive those beautiful structures which we admire so much. Except a very few, all live by the destruction of others, and have to be continually trying to circumvent their neighbours to escape extermination. How exactly they are fitted to contend with adverse circumstances is shown by the fact that so many survive, and although in the long ages which have passed since life first originated on the earth, the destruction has undoubtedly been enormous, it must have been always of little importance in comparison with the survivals.

This is beautifully exemplified in a thousand

ways. Here in the forest we see evidence of enormous developments having taken place in the past, and, what is of far greater importance, actually in progress before our eyes. Some objectors to evolution have gone so far as to state that variation is mainly due to man's interference; but when it is considered that cultivation is applied for the purpose of perpetuating certain characters, and developing them at the expense of others, we see at once this cannot be true. Even the original divergences which he utilises took place entirely apart from his influence, and however he may try to produce certain changes, he can never succeed unless the initiative has already been taken.

The examples we have just given are illustrations of the more active side of the great struggle, but a thousand others might be quoted where plants at first sight appear almost passive. Yet even here a grand work is always in progress, in every case more or less connected with the interdependence between one life and another. Without the tree the epiphyte or parasite could not exist as such, without the flower the bee would be starved, and without the numerous fertilising agents most plants would be unable to produce seeds. In temperate climates the woods are made up of two or three species—sometimes of only one. These blossom almost simultaneously, and their pollen floats upon the wind to long distances. Here in the tropics, however, things are different—the wind has little power over such a mass of foliage. Whether looked upon from above or examined carefully from beneath the undulating roof seems perfectly still. The branches are rigid, the leaves stiff, and even the flowers thick in texture. There is nothing comparable with the birch or beech, much less the trembling aspen. The changes produced by a sudden gust upon these trees are entirely unknown here, as are also those beautiful effects of light and shade which delight the painter and poet. It follows also that such yellow clouds of pollen as hover round the pines are entirely wanting, there not being, as far as can be seen, a single wind-fertilised tree in all the Guiana forest.

Being unable to utilise the air currents the flowers have had to look round for efficient substitutes, and these are found in the host of insects which hang in clouds over the forest canopy at night and buzz around in swarms during the day. Even birds are utilised by some of the larger flowers, the pollen collecting on the bristles at the root of their beaks. Without living helpmates many a tree would become extinct, therefore every effort is put forth to attract and induce winged creatures to render this assistance. The principal means to this end are colours and perfumes, the former for diurnal and the latter for nocturnal insects. Brilliant colours are of themselves sufficient to attract butterflies and bees—lurid and dull tints are usually accompanied by odours more or less disagreeable to our senses but pleasant to flies. It might be suggested that in the one case the flowers are gaudy because they are open to the fierce rays of the tropical sun, while the others are the contrary, on account of their blooming in the shade. But, with the apparent capability of choosing between the ends of the branches, the axils of the leaves, and the trunk, all attain their object, whichever place they have selected. Here and there in the forest we come upon a tree the flower stems of which originate below the canopy of leaves, and this appears so strange that we wonder why such exceptions occur and what particular advantage is derived from this position. Except that here they are more in the way of the shade-loving insects we know of no other reason, and must at least accept this provisionally. The cacao is an interesting example, and it can hardly be considered as having degenerated, but rather as being most beautifully fitted to its environment.

It is hardly necessary to say much of the handsome flowers that bask in the sun and derive their glowing colours from his rays. We may, however, mention those which open early in the morning and close before the light becomes too strong for their delicate complexions. Perhaps the best-known example is the morning glory, but other Convolvulaceæ, orchids, and bignonias are equally striking. Then, again, there are species which cannot bear the least dampness, but wait until the sun is well up before they open, and close again as the shadows lengthen or earlier in cloudy weather. Another class open towards nightfall, but these approach that very large division which is nocturnal in its habits.

What a glorious assemblage are the night bloomers! The Victoria regia, the Cereus and other Cacti, the Daturas, with a host of other white flowers, all open at sunset, distil their perfumes and diffuse them through the damp air. Then come those other white flowers which remain open during the day, but continue almost, or quite, scentless until recovered from the effects of the burning sun at nightfall. In the forest many trees bear insignificant-looking greenish-white flowers, hardly worth looking at. But, paddle along near the bank of the river at night, and you will be astonished at the different perfumes wafted from these poor little blossoms. Now, it is an odour of the jasmine type, so strong as to be quite sickly, and anon one of those half-aromatic scents which almost exhilarates. Where they come from or to what tree they belong it is impossible to say— perhaps they are being distilled by a tree close to the bank, or it may be by one of which we know nothing, a quarter of a mile away.

Perhaps the most interesting point in connection with these perfumes is that they are distilled at certain times, and then only for short periods. Sometimes the flower opens, carries on its work for an hour or two, and then closes, either altogether, or in a few cases, to repeat the process at the same hour next day. These latter are, however, the exceptions; as a rule the flowers remain open, but apparently quiescent, until the time comes round to resume work. This is beautifully exemplified in certain orchids, and sometimes causes great disappointment to their owners. Perhaps you have been telling a friend of the delicious perfume of Burlingtonia fragrans; that it brought to your memory a pretty lane in the old country, where as a child you used to go and smell out the sweet violets when they were so hidden among the leaves that you could hardly find them otherwise. "Come along!" you say, and take him to the flower. "Is it not delicious? Does it not bring up the old memories? What a close resemblance!" Your enthusiasm carries you away, and you expect an eager response. But no, not a word does he say, although his politeness prevents him from contradicting you. The fact is the flower has already done its work for the day, and not the slightest trace of perfume lingers on its beautiful petals.

This is by no means uncommon, in fact, intermittent distillation is almost general in the white flowers of the tropics. Even in temperate climates there is a striking difference between the amount of perfume given out at different times, but rarely are the flowers quite scentless at any time. It might be thought at first that these alternations were erratic, but close observation shows that they are nearly true to the minute, and if carefully timed would almost certainly be found to coincide with the period when the fertilising agent is on the wing. For this is the simple explanation; the flower can only be fertilised by a particular kind of insect, and all its efforts are put forth when that insect is likely to be hovering around. Without the flowers the bee could not exist, and without the bee no seed would be produced. That this interdependence is complete can be easily proved in our gardens, as indeed it is also patent to every one who has a conservatory in other countries. Without some particular insect the plant flowers over and over again without producing a single perfect seed, and as this agent does not exist away from its native habitat the plant remains barren.

That there should be such a nice arrangement seems more than mere coincidence, and even shows design in the plant rather than the animal. We may perhaps say that the insect fixes its dinner-hour at the time when its food is ready, but in the cases of many perfect insects their life is so short that preparation would have to be made in the pupal stage. It is therefore easier to suppose that the first choice was made by the plant. The interdependence is, however, so very close that it is just as hard to decide in favour of one as of the other.

The shapes of flowers, like their perfumes and colours, also show wonderful adaptations of means to a particular end. The great tubular flowers are just deep enough to allow the long proboscis of the humble bee or moth to explore their depths. Again, we are met with the question, was the contrivance of the flower or the bee adopted first? We can conceive that in some past age this deep funnel was but shallow; probably made up of fine separate petals and as wide open as a buttercup.

Bees kept up a continual irritation in their search for nectar, and caused the sensitive stigma to recoil as it were and bring its petals closer together. Only the bees with the longest proboscis could now reach the reservoir, but there were still enough of these to keep up the irritation. As generation after generation passed, the deepening and elongation still continued, until now we have flowers with tubes six inches or more in depth, and humble bees, the proboscids of which are able to sip nectar from the very deepest, at the same time performing the necessary work of fertilisation.

A great deal more might be said of flowers, but we are unable to deal with other than a few of their contrivances, and must proceed to the fruit. Here also the interdependence of plants and animals is beautifully exemplified. But, instead of insects, we now have to deal with birds, bats, rodents, monkeys, and even fishes. The fruits of the tropics have succeeded in protecting themselves against insects, and are consequently never subject to the attacks of wasps like their cousins of more temperate climes. It is true that beetles make sad havoc among them, but otherwise, their thick skins are impenetrable to anything smaller than birds or bats. The orange family seems to have even gone farther, and by means of bitter secretions and stinging essential oils kept off all comers. Other fruits have not gone so far, although many of them try their best, with thick skins impregnated with tannin, and in some cases poisonous milky juices, to preserve their edible seeds from destruction. Few, however, succeed in this—the animals have learnt to peel them and get their share.

On the other hand, a large class of fruits are obviously intended to attract. It is not to the advantage of a tree that its seed shall lie round the trunk and be lost in the struggle for existence. It has therefore followed that efforts have been made to provide something to feed the animals, but at the same time to protect the life-germ from extinction. This is often done by covering the seed with a hard shell, outside of which comes the luscious pulp which forms the attraction. Birds sometimes carry food to long distances, to eat at leisure or to feed the young, and thus the seeds are dispersed in every direction. Sometimes they are eaten and ejected without injury, in fact, rather with benefit as far as can be judged—most of the figs and loranths are examples of this. Where there are bats their lurking places are often strewed with fruit, more or less disfigured on the outside, but never really injured as far as the germinating power is concerned. These little creatures hang in large numbers about the festoons of creepers which border the creeks or inside hollow trees, and do a great deal towards dispersing the seeds by bringing them within reach of the flood. As nearly all of them float they are carried down the stream and scattered in a thousand places which otherwise they could never reach, even finding their way to the shores of Africa, Portugal, and England.

Without the flowers and fruit the forest would be entirely deprived of animal life, and in the absence of the clouds of insects not a seed could be perfected. The interdependence is so close that they appear as if striving with each other to see which can get ahead. Trees secrete poisonous juices to keep off larvæ, but the caterpillars do not mind that. They only develop a greater power of digestion and assimilation so as to be able to thrive on the poison. However dry and harsh, thick and leathery, hairy, scaly or bristly, the leaves may be, some insect can be found capable of devouring them without injury to itself. It would never do for a plant to lay itself open altogether to their attacks, as then its extinction would be certain, therefore it tries its best to ward them off. Those who do not put on armour generally protect themselves by acrid, bitter and poisonous secretions. No doubt they succeed to a certain extent, but it always happens that some of their foes still get the better of them.

Finally, the geological formation of a district has considerable influence on the distribution of life. Apart from the obvious fact that a desert must be almost devoid of life, there are differences between hill and dale, sandstone and granite, and clay and gravel, which result in corresponding variations in the fauna and flora. In British Guiana some have gone so far as to say that they can tell when an auriferous district has been reached by the prevalence of certain kinds of birds and monkeys. This can be easily understood when the close connection of the trees with the soil, and the fruit with the animals, is considered. As in the case of civilised countries the number of inhabitants depends a great deal on the food supply, so here the denizens of the forest vary from the same cause.

V.

THE STRUGGLE FOR LIFE.

IN the forest we are among a great crowd, compared with which the largest assemblages of people are infinitely little. There is absolutely no room for a single addition until one of the number succumbs to the continual strain, or is conquered in the fight. Standing room is all they require, but few succeed in obtaining enough space to stretch their arms. Above our heads is a verdant roof which shuts out every ray of sunlight from beneath, and we wander among the columns of this great structure as if it were the mazes of some interminable cave. The columns are enormous brown pillars or stalactites, stretching from floor to roof, their resemblance to the work of an architect being increased by the almost universal provision of buttresses.

All are on the same level. No individual can afford to let another get above him. Every one is straining upwards to obtain a little more sunlight, but there are no laggards. It is rare indeed for a giant mora or silk-cotton tree to rise above its fellows, so that looked at from above the surface is a uniform level—a verdant plain, undulating with the ground on which it stands. Thousands of little streams flow through its arcades without breaking the continuity, but here and there a great river produces the effect of an embankment, as the forest slopes down into the water.

To the casual observer inside the forest everything appears almost lifeless. These gigantic buttressed pillars are emblems of strength and power, but they do not seem to utilise their capabilities. Look a little closer, however, and a flood of light is thrown upon the matter. Everything in nature has its purpose, and we may be quite sure that the force and energy stored up in the giants of the forest are there for use. And that they are used is equally evident on patient investigation. Every one of these trees is a living creature straining to hold its own in the battle for life, and continually struggling with its fellows for the mastery. Under our feet are interlacing roots filling up every inch of ground, and above our heads expanses of leaves to prevent a single ray of light being wasted. These represent the labour of many years—positions gained by unremittent effort. Thousands of competitors have been overcome in the struggle, and here we have the strongest and fittest surviving to continue the strife. For it does not follow that because they have attained a certain position, they are, therefore, the more able to take things quietly. They are surrounded by neighbours, every one of which is prepared to take advantage of the least opportunity. They spread their branches over each other, push through any little aperture where light has been allowed to penetrate, doing their level best to smother the weak ones. Fortunately, however, every tree is continually on the watch, and, as his neighbour raises himself a little, he extends a branch upward to cope with the enemy. Down below a similar struggle is going on. In the same way that the branches strive to gain a little more sunlight so the roots strain after food and water. They stretch to long distances, here climbing over or under the great feet of their rivals, and there disputing every drop and morsel with their delicate fibrous mouths. Above, as well as below, the immediate surroundings are occupied by dense assemblages of branches, through which it is impossible to penetrate, but in both places there are always extensions to long distances. These wind in and out for hundreds of feet, always looking out for the necessaries of life and for material with which to carry on the struggle.

Every living thing, whether plant or animal, has to work for its living. The Utopian idea that it might be possible for man to live without work is scouted entirely when we come to study things as they are. Not only is it necessary to labour, but to fight as well. There are always more mouths than bread to fill them. The weakest go to the wall and the strongest survive everywhere. If anything happens to put a species in the background it becomes extinct and the others ride roughshod over it. As with the species so with the individual — millions are born only to die before they attain maturity. The forest giant works for many years before he attains a position to produce offspring and how rarely does one of his progeny succeed. He scatters his fruit over the ground by tens of thousands, gives them manifold provisions for preserving the precious germs of life and for securing their distribution, and then leaves them to do the best they can. If they try to germinate under his wing they are ruthlessly smothered. He does not intend that they shall be *his* rivals—they must look for a vacancy, and if there is none go the way of the millions of others whose parents have also apparently strained every effort and spent their strength for naught.

Poor little creatures, how pitiful it seems that they should die in infancy! We see them everywhere in the forest. Here is a great heap of seeds, every one with its white cotyledons pushing through the smooth brown coat, which has been its swaddling cloth for a few weeks or months. Rain has fallen and caused the germs to swell, and now the little ones are beginning to look around them and strain for the light as their fathers have done. But what is the use? Here they are in this dark cave, and not a single ray of the glorious tropical sun can reach them. The cotyledons push out and remain colourless, or perhaps a leaf or two may be produced on a long, weak stalk, but without light they can do nothing. One after another withers and dies, and the next dry season finishes off the whole. Perhaps a few favoured ones may have gone a little farther. The place where they were carried by some running stream is not quite so gloomy as the recesses of the forest. A hundred little trees have gained a footing and are now making every possible effort to get upwards. Here they are, as close together as they can pack, long slender stems, no bigger than your finger, the tallest twenty feet high. The struggle has commenced and the vanquished are easily distinguishable. Here is one a foot high, and scattered between the victors, others of different sizes. All have lost their leaves, and although some might still put out fresh ones if they had an opportunity, it is easily seen that all are doomed. They have lost their place already and are out of the running.

And, what of the victors? They look vigorous —you can almost see them growing. Their upper leaves are all on a level and glow with the most beautiful tints. There are no branches—they cannot afford to waste time and energy in putting out side shoots which would be useless. Their motto is "Excelsior"—higher and higher still, and woe betide the hindermost. There is room and light enough for perhaps two trees, and here are a hundred. They cannot all succeed. Look closely and you will see that some are more sturdy than others; they are not quite so crowded and their roots collect more nourishment. These will soon be ahead, but still there will be too many. The struggle will become more and more intense the nearer they get to the top, and when the few survivors begin to throw out branches every one will elbow his neighbour until only the favoured ones remain. As for the others they will be represented by a few dry sticks, which soon crumble into food for their successful rivals.

Having got so far it might be supposed that the victors would rest upon their laurels. But no, they have only attained to manhood and must now fight with their equals. The elbowing is carried on above just as it was on a lower level—every effort is strained to prevent a new-comer getting his share of the sunlight. While the struggle was going on below the trees have been stretching their limbs over the little opening and only a slit may remain through which the sturdy youth has pushed his way. He naturally extends himself and tries to rise above the others, but they, in their turn, push through his branches until all are interlocked one with another. So great is the confusion on this account that it is impossible to discern from underneath what leaves belong to a particular trunk.

If this were all that the forest giant had to contend with it might be considered as a fair game, but, unfortunately, the tree has other enemies. Nowhere in the world are there such enormous climbing and scrambling plants as in the South American forest, and all these are more or less dangerous. Like the trees they aim to get upward to procure enough light for their gorgeous blossoms, but, having such limp stems, can only do so by means of their victims. Some twine round the trees like monster pythons, others produce aerial roots which cling to the bark; some push themselves through the branches and twigs, and then spread out their arms as it were to prevent falling back, while one at least walks up the trunks like a cat by digging its claws into them. As their seeds germinate in the forest they are poor, weakly, little things, seeming altogether helpless, only looking for some friendly trunk to help them in their progress upwards. The tree is unfortunately wanting in the ability to refuse this help and consequently the twining malefactor obtains a footing. It grows very quickly, its soft body accommodating itself to the shape of its host, and coiling round and round until it reaches the top. Here its point is pushed through the smallest crevice and the monster opens its leaves to the sunlight. What a striking change! Branches are thrown out in every direction, the coil thickens and compresses the tree-trunk, and very soon the poor forest giant begins to suffer. Being an exogen it cannot bear compression—its leaves begin to fall, and this makes all the more room for the great mat which is being spread all over its head. By and by, as the vegetable python constricts its host more and more, the tree dies; then the victor covers it with a flowery pall, spreads over a dozen other trees, and glories in its triumph. Later, when atmospheric agencies and termites have done their work, we see the empty coil of the monster hanging in mid-air and helping with others to produce the effect of great cables hanging from tall masts.

Other bush-ropes (as they are called in Guiana) do not constrict, but simply smother their hosts. As they reach the sunlight they, by growing faster than the tree, quickly spread over the neighbourhood until the light is entirely shut out from below. Of course the tree does its best, pushes a twig here and there through the mat of its enemy, but rarely if ever conquers him. For ages the climber has been at work and learnt how to fight for its position, until now the tree is almost entirely at its mercy.

However, even this powerful monster is afflicted with some disabilities. Like other plants its seeds must have a certain amount of light before they can even begin to grow, and if the canopy above is quite unbroken its efforts to rise are fruitless. But there is a class of plants which has got over this difficulty by commencing at the top. Woe betide the forest giant when he falls into the clutches of the clusia or fig. Its seed being provided with a pulp, which is very pleasant to

CLUSIA FLOURISHING ON LOCAL TREE.

the taste of a great number of birds, is carried from tree to tree and deposited on the branches. Here it germinates, the leafy stem rising upward and the roots flowing as it were down the trunk until they reach the soil. At first these aerial roots are soft and delicate, with apparently no more power for evil than so many small streams of pitch, which they resemble in their slowly flowing motion downward. Here and there they branch, especially if an obstruction is met with, when the stream either changes its course or divides to right and left. Meanwhile, leafy branches have been developed, which push themselves through the canopy above and get into the light, where their growth is enormously accelerated. As this takes place the roots have generally reached the ground and begun to draw sustenance from below to strengthen the whole plant. Then comes a wonderful development. The hitherto soft aerial roots begin to harden and spread wider and wider, throwing out side branches which flow into and amalgamate with each other until the whole tree-trunk is bound with a series of irregular living hoops.

The strangler is now ready for its deadly work. The forest giant, like all exogens, must have room to increase in girth, and here he is bound by cords which are stronger than iron bands. Like an athlete he tries to expand and burst his fetters, and if they were rigid he might succeed. But the strangler is like a python, and almost seems as if provided with muscles. The bark between every interlacing bulges out and even tries to overlap, but the monster has taken every precaution against this by making its bands very numerous and wide. We can almost see the struggle, and, knowing what will be the result, must pity the victim.

As the tree becomes weaker its leaves begin to fall, and this gives more room for its foe. Soon the strangler expands itself into a great bush, almost as large as the mass of branches and foliage it has effaced. Its glossy leaves shine in the sunlight, and it seems to glory in its work. Every branch is clean and sleek, not a lichen or fungus can find shelter anywhere. It has got on the shoulders of the forest giant, but does not intend to support in its turn even the tiniest dwarf. If we could forget its murderous work, how we should admire it! Take the *Clusia insignis*, for example. Here we have one of the most beautiful shrubs in the world. Its thick leathery leaves shine as if polished, and its green sleek branches always look clean and healthy. As it sits crowing, as it were, over its victim, the contrast between them is most striking.

CONFUSED ASSEMBLAGE OF AËRIAL ROOTS OF A CLUSIA.

Perhaps the forest giant is dying—the few leaves remaining are yellow and sickly. No flowers have been produced for two or three seasons, and even the branches look shrivelled. There is not the least hope of recovery: it only remains, therefore, to submit to the inevitable, to die and give place to the strangler.

How pitiful these victims appear! Sometimes in passing along a creek they are to be seen here and there in all stages. Now and again the clusia or fig has been content with the destruction of one branch, and instead of fastening a network of fetters round the trunk, runs down one side. In such cases only half the tree dies, and the remainder looks as if in the loving embrace of a friend. Those which have been more unfortunate are yet standing. Here is one with a dark mass hidden in its canopy of foliage. As yet the murderous work is only beginning, and no serious mischief has been done. Further along, however, is another, which is obviously suffering. Its leaves are so few, that the green mass of the strangler dominates and helps to cover its almost naked limbs. Now we come upon another quite dead. Where it is not enclosed in the living fetters, the bark hangs down in great flakes, while branches depend from the festoons of bush-ropes which help to hide the unsightly object.

Termites are already at work, as may be seen from their great black nests which occupy every fork, and beetles are found everywhere above and below. Woodpeckers hunt for larvæ, making great holes in the rotten timber, and parrots build their nests in these excavations. Water trickles down into them and helps on the work of destruction, which, although it may take a few years to complete, is nevertheless done more quickly than might be supposed. A mora pillar may last from fifty to a hundred years in a building, but hardly a tenth as long in the forest. If we look carefully around us, we see examples of entire obliteration, a clusia or fig standing on its reticulated hollow pillar with only a heap of brown humus at its base to show what has become of the trunk which once stood up in all its majesty on that spot.

As if the trees had not enough to contend with from elbowing, smothering, and strangling, another enemy, the blood-sucker, has arisen to disturb their peace. Species of Loranthaceæ—the mistletoe family—are very common in the forest. Unlike the pretty Christmas bush, however, these are monsters of a most pronounced type, often forming bushes twenty feet through. Like the stranglers they are propagated by birds, which eat the glutinous pulp of their berries and clear away the

WILD FIG, AFTER DESTROYING ITS HOST.

seeds by rubbing them on the branches of trees, to which they adhere. Myriads of these seeds germinate in every direction, on leaves and even rotten wood, but if they do not happen to find a congenial spot on some living branch or twig, they perish very quickly. Instead of a radicle the germ produces a sucking disk, which immediately adheres to its living support and begins draining the sap to support, first its cotyledons, and then its long whip-like stems. As the parasite gets strong, its long extensions spread from branch to branch, and from twig to twig, everywhere extending octopus-like arms provided with sucking disks, which adhere to and bleed the tree in a hundred different places. Branch after branch is dried up, but as the loranth has many strings to his bow, this does not hurt him much. There are always more to conquer, and unless the tree stands alone, which is of course impossible in the forest, he rarely comes to grief. It is not to his advantage that the tree should die quickly, and therefore the longer it can support him the better. However, even the most sturdy giant of the forest suffers greatly from such continual depletion, and may be so weakened as to lag behind in the race for life, with the ultimate result that it is smothered by its fellows.

Surrounded by so many rivals of its own class,

it naturally follows that the forest giant puts on defensive armour. Perhaps the best examples of this are found in the palms. As endogens they have certain advantages which are denied to members of the other great division of the vegetable kingdom. Having no branches they rise straight upwards without difficulty, and their pointed leaf-buds pierce through the thickest canopy. Then, every exposed part of the crown is particularly stiff and hard, so that it has hardly a weak point. Not content with this, many species have put on spiked armour, which effectually prevents any weak seedling from growing in their neighbourhood, and tears them to pieces if already there. Such an array of needles as is borne by some of the species of Bactris, is simply appalling. Stem, midrib of leaf, and even the leaves themselves, are so beset with them, that not even man can penetrate such thickets as they form. Not content with a single stem, like most other palms, they throw up one circle beyond another, cover a considerable area, and conquer everything that stands in the way.

Palms are also well protected against stranglers and blood-suckers. A fig will make use of the tall column and throw its aerial roots down it to the earth, but rarely does it gain anything like a position. The crown of the palm overshadows its leaves, and however it may constrict the hard stem, not the slightest injury results. The magnificent Oreodoxa oleracea—the palmiste of the West Indies—is often covered with a network of the creeping fig (*Ficus stipulata*), which gives to what otherwise would have been a bare column, a most pleasing decoration. This, however, can only be produced when the palm stands apart from other trees, as in the forest sufficient light cannot be obtained. Possibly the high position in the vegetable kingdom which has been obtained by palms is partly due to their perfect immunity from the attacks of stranglers and blood-suckers, as well as to their greater ability to hold their own in the great struggle for life. As for loranths they can never succeed in getting a foot-hold on the hard palm fronds, and if it were not for insect enemies these "princes of the vegetable kingdom" would be perfectly safe.

We have already hinted that the dense shade of the forest is a protection against undue rivalry. Only light is wanted to make the space between the tree-trunks an impenetrable jungle. In temperate climates most trees are bare for half the year, but fortunately for them this happens when nature is sleeping. Here in the tropics there is no rest. Day after day, and to a certain extent even at night, throughout the whole year, the trees are continually at work. They drop their leaves, generally twice in the twelve months, although a few only do this once. When the change takes place it has to be got over hurriedly. It would never do to have the branches and twigs bare for even a few weeks with so many rivals prepared to take advantage of such an opening. If there were an uniform resting period the disability might be less, but as it is every individual differs more or less from others, even of the same species, and chooses its own time. From January to March, and again six months later, fruits ripen, the leaves fall off one by one, and almost immediately—perhaps in a few hours—the tree is again clothed with a vesture far more beautiful than that which it has just cast off. If this change were at all uniform, the forest would at this time present a most gorgeous appearance, to which the autumn colouring of North American forests bears no comparison. Even a single tree is "a thing of beauty," but like that quality in many other things soon gives place to mere uniformity. As the buds expand, the flaccid leaves are seen tinted in most inimitable shades. On backgrounds of pale green are painted with the delicate hand of nature rich crimsons, browns, olives, yellows, and whites, which glow in the sunlight and tinge the rays which for a few hours are enabled to penetrate the otherwise dense canopy.

In that portion of the forest which covers the lower grounds near the coast, the struggle is most intense, on account of the richness of the soil and the plenitude of water. On the sand-reefs, rocks, and mountains, however, plant-food is more difficult to obtain, and here the strain for bare existence is so great that a longer rest seems necessary. Instead, therefore, of performing two tasks a year, the trees generally have but one flowering and one fruiting season. Again, as little rain falls for the three months from September to November, that time is chosen for a partial rest. This, however, is not like the hibernation of plants in temperate climates, but simply a less active condition where the strain is slightly relaxed. The leaves fall one after another, but almost before you can appreciate the fact that the tree is bare, new buds open all at once and work is resumed.

That they should require a little rest at night is only to be expected. Even in other climes the sleep of plants has been noticed, but nowhere is it so strikingly exemplified as in the tropical forest. Diurnal flowers droop, leaves fold together or hang as if exhausted, and even twigs lose some of their stiffness. All, however, do not choose their hours of relaxation at the same time, but as in the case of animals, there are some which work at night, when they bring to perfection that intricate process, the fertilisation of their flowers. The night workers close their corollas against the bright sunlight, and only open them to commence the work of distilling perfumes after nightfall. Then the whole forest is noisy with the din of myriads of insects, and perfumes are wafted in every direction, floating downward from the tree-tops, and rising from the host of water-lilies which cover the open parts of the creeks.

At such times (the flowering seasons) every plant must be particularly active, as its leaves work during the day and its flowers at night. The most strenuous efforts seem to be put forth to attract nocturnal insects, with most wonderful results, and when it is considered that during all this time the tree continues to wage war against its neighbours so as to hold its own, the amount of labour is seen to be something enormous. For years it has been gathering strength and struggling for room to develop its flowers, and now the critical period has arrived.

CREEK SCENE.

VI.

ON THE RIVERS AND CREEKS.

IF the struggle for life is so intense inside the forest, how much greater is it along the banks of the rivers and creeks. As we have before stated, the narrower streams meander through dark arcades, and only when they are over fifty feet broad do we get anything like a break in the continuity. Even then it is only here and there, where a tree has been undermined by the flood, that anything like a patch of sunlight comes down to the level of the water.

In passing along the banks of the great rivers of Guiana, we come upon openings, from which flow swift streams of coffee-coloured water. To the casual observer they are little bays hollowed out of the bank, and backed by masses of vegetation which effectually hide everything beyond. These are the mouths of tributary streams, called creeks, which drain the forest, feed the rivers, and through them tinge the ocean for over fifty miles from the shore. If we paddle our bateau into the entrance, it is soon found that the appearance of a *cul-de-sac* is caused by an abrupt turning, on rounding which the river is shut from view, and a real fairyland disclosed.

Painter and poet have depicted the brooks and small rivers of temperate climates, but all their glorification of nature seems tame when applied to a creek. Even the ordinary observer becomes enthusiastic, while the naturalist experiences a feeling of ecstacy that is simply indescribable. The fatigue of a long boat journey on the open river, where the fierce rays of the tropical sun poured down incessantly and blistered his face, neck, and hands, is all forgotten, and he can do nothing but sit up and feast on the beautiful. Every bend brings up a new scene. Here is a great mora towering to a height of a hundred and fifty feet, from which hang festoons of creepers decorated with large flowers of most gorgeous colours. Below and in the foreground are a thicket of tree ferns, great clumps of marantas and heliconias—a hundred species of shrubs and low trees. A little farther we come upon reaches where the most striking objects are palms; here a troolie with almost undivided leaves twelve feet long, farther on a clump of the graceful manicole, and in another place perhaps the stately fan-leaved eta. Now the creek is almost closed by a lattice of bush-ropes, and then we have to pass under a leaning trunk or branch almost touching the water. Hundreds of cord-like aerial roots depend from the topmost branches of the trees, and have to be moved aside as we get amongst them, while great bunches of flowers depend from the creepers, which also obstruct the way in some places.

If the creek has not been kept open by Indians, it is often choked with vegetation. A dense wall of creepers forms a curtain, and we can only push through by aid of our cutlasses, which are always carried for this purpose in bush travelling. Under water are the remains of trees which have fallen during several centuries. When the water is low we see them lying in inextricable confusion on the bottom, and every now and then our bateau grazes one that stands higher than the rest, or perhaps lodges upon it until pushed off. Here a great tree has fallen and blocked the way altogether, and we must use the axe if we desire to proceed farther. With a light canoe we may perhaps get out on the trunk and haul the frail craft over, or if this is not possible clear away the branches sufficiently for a passage. Now and again it is possible to push the craft under, while we scramble over the trunk.

These fallen trees give opportunities to seedlings not to be had in the forest proper. Floods have loosened their roots, and they come down some day with a crash, allowing the sunlight to penetrate where there are myriads of seeds only waiting its advent to enter the field of battle. With our knowledge of the intense struggle always going on in the forest, we can hardly conceive how a seedling can make headway at all on the banks of the creeks, where the strife is so much greater. Its rivals are so many, and their tactics so different, that success would almost seem impossible. If we look at the lower vegetation, we find that a great many species have acquired the power of increasing by suckers, which faculty is wanting in the forest trees. Marantas, heliconias, palms, and arums, spread over large surfaces by this means; every new sucker is protected by the parent clump, and pushes its rivals farther and farther as it increases in size. Here also are dense shrubs with very stiff branches, in many cases armed with spines. Between these and the tender-leaved marantas there is a continual struggle. The latter try their best to cover everything with their glorious crowns, but the thorns and stiff twigs continually tear their great leaves into ribbons, and refuse to be smothered. Some have spines only in their early stages, and lose them as they become older, when their positions are better assured. Even the tree-ferns are armed in such a manner, that they all deserve the name *ferox*, which has been given to a species of Alsophila.

Creepers that twine are necessarily repelled by thorns, and probably for that reason are comparatively rare. Most of the smothering host which comes to the front on the banks of the creek are scramblers. They, as it were, crawl over everything, holding on by contrivances which are particularly ingenious. Some push their small shoots through the dense bushes, and then, as a boy in climbing a tree, puts his arms over the branches, they spread out stiff hooks with the points downwards to prevent their slipping back. Others, like Bignonia unguis, can take hold between the spines and keep their leaves at a safe distance. Once these scramblers get a footing there is hardly a limit to their extension. Long whip-like branches spread in every direction, appearing as if continually on the look-out for a resting-place. They pass over thorny bushes and other uncongenial spots, and only when a tree suits them do they hasten to cover it with their flowery pall.

Here and there a little bay at the turning of the creek is flooded with sunlight, and this makes the struggle all the more intense. If the current is not very strong a bed of *Cabomba aquatica*, with its pretty shields, will occupy the space, and beyond it a clump of white lilies (*Crinum*). Then comes the really impenetrable jungle, which is so dense, that not a single ray of sunlight reaches the ground. But there are few open places in the narrower creeks. For long distances the trees meet overhead, and make everything beneath almost as dark as in the forest. If, however, the smallest ray of light can penetrate, we see a struggle going on below for its possession. Again, for miles the stream flows through a living tunnel, and here nothing whatever can grow; the tangle of branches above our heads is free from epiphytes, and even leaves.

We have not mentioned the epiphytes, as they are important enough to deserve a special chapter. We may, however, remark that it is in the creeks where the lovely orchids sit on branches above our heads, and, together with ferns, peperomias, cacti, bromeliaceæ, and aroids, decorate every branch and twig. As if looking down from above on the intense struggle, these seem like spectators watching the fight, but having no interest in it. If they fell among the crowd they would be quickly suffocated, but as their aerial roots cling tightly there is little risk of such a catastrophe.

We come now to the great rivers ; their banks are too uniform to be picturesque. The forest trees come as near to the banks as they dare, and then stop, allowing a crowd of prickly shrubs to extend themselves into the ooze. Looked at from the river, the green expanse seems to rise from a verdant bank, as if the shore were far above the water, instead of being nearly on the same level. The tall trees cannot hold their own in the mud, therefore they give place to a different type, which has little or no trunk, and sits down as it were to anchor itself by means of special contrivances. Several species of Leguminosæ, including Drepanocarpus lunatus, Muellera moniliformis, and Hecastophyllum Brownii, form dense thickets, and extend as far from the bank as they dare. In front of these is an advance guard of mocca-mocca (*Montrichardia arborescens*), which is as it were drawn up in rank to keep back the flood. Growing in the water, this monster arum develops great club-like stems, which come up as close to each other as they can pack, and rise like rows of palisades to the height of twelve feet or more above the surface. As if this were not a sufficient encroachment on the open space, the floating-island grass (*Panicum elephantipes*) anchors itself to the mocca-mocca or bushes, and extends just as far across as the rapid current will allow. In dry weather, when the water is low, and the stream has little power, the extensions from either side meet in the centre, and close the passage-way for a time, only, however, to be torn away in great masses as the floods come. At such times great patches, fifty feet or more in diameter, are seen floating down-stream, sometimes carrying with them monster camoudies (*Boa murina*), or other snakes. As these masses are caught by the sea-waves, they are thrown back upon the beach, where they lie in great heaps until, dying, they go to help make up that extension of the coast-line which is continually driving back the waves to a greater distance. Sometimes a great tree, whose timber is light enough to float, gets entangled in the grass, and becomes the nucleus of an immense raft, which is continually increasing in size as it gathers up everything that comes floating down the river. The grass extends over the whole mass, and mats it together until a formidable obstacle is produced, but notwithstanding all its efforts the dam is imperfect. When eight or ten inches of rain fall in a day, and the river rises sixteen to twenty feet, the barrier must go. However it may be attached to the bottom by a thousand anchors, it has to give way when the rise takes place, and here the hollow stems of the grass help in its own destruction. By their numbers they act as buoys, drag the great tangle of trees and bushes to the surface, unloose their own anchorage until the mass sails away, ever on and on, to be broken in pieces and dashed on the shore, or perhaps carried far out to sea.

Thousands of floating trees and patches of grass are carried down by every flood, and are not uncommonly found out of sight of land. Sometimes a great tree resembles a wreck, and ships have been known to steer out of their courses and send boats to see if anything living could be found in the tangle of presumed spars and rigging. As may be supposed, such a mass as we have described often plays sad havoc with the river banks along its course. Now it is driven against the shore on one side, and carries away a clump of mocca-moccas, and farther on sweeps a little thicket before it ; now a part of the mass gets entangled in the props of a mangrove, and for a short time the whole is brought to a stand-still. Something is bound to give way—either the mangrove is dragged from its anchorage in the mud, or the floating island must part with some of its constituents. The flood is inexorable; like the policeman, it tells the monster obstructionist to move on, and there is no possibility of disobedience.

However, it sometimes happens that an obstruction of a particularly obstinate nature succeeds in withstanding the flood. Perhaps it has existed long enough to allow the climbers on the overhanging bushes to lay hold, and the mocca-moccas to grow up and penetrate it like stakes. It was stranded in a shallow, and has been the cause of further deposits on and around it. By and by down comes the flood, bringing with it great masses of similar material, most of which is deposited on the fast-forming bank, until it becomes an impregnable barricade, narrowing the channel to a considerable extent. The waters become higher, and the current swifter. Something has to go, but it is not this late erection. The opposite bank is undermined, one bush after another goes down with the flood, trees fall over and are also carried away, and a few months later the great river has a new bend. This sort of thing is continually going on; now the stream overcomes the plants, and anon they are borne away to the ocean, the general result being, however, that everywhere except where the bed is rocky the channel is quite unstable.

It happens sometimes also, though rarely, that the obstruction takes place in the middle of the stream. A very large tree is carried along and deposited in a comparatively shallow place, where it settles down and becomes the nucleus for a deposit of *débris* and silt. Then the floating grass anchors itself to some of the projecting branches, and spreads all over until a little island is formed. Layer upon layer of mud is arrested at the upper end until there is only a foot or two of water, and on this bank some of the mocca-mocca roots, that are continually coming down, get stranded and commence to grow. A year or two after a dense living palisade protects the small island, prevents the silt from washing away, and helps to increase the deposit by keeping a little still water behind. Seeds come floating down in myriads, germinate in the tangle, grow into great shrubs, and continually gather more mud and *débris* until the little island lengthens down-stream to a considerable distance. The semicircle of gigantic arums is sufficiently elastic to bend before the weight of water; the plants may be torn up by the roots, but never broken off. A thousand little eddies make them tremble, and move backward and forward as if alive, but still they go on producing one shoot after another, and crowning all with their handsome arrow-shaped leaves and waxy spathes like great arum lilies. However the floating grass may try to cover them with its dense mat they push through the thickest covering. Each shoot is a living spear, ready to pierce almost any accumulation, and rise well above the surface before opening its leaves to exclude the light from below. Not only is it pointed, but also armed with thick short thorns, which tear the leaves and stems of everything the flood brings in its way.

The great rivers of Guiana all contain islands of different sizes, some as many as ten miles long, and it may be confidently stated that nearly all have been built up in this way by means of the mocca-mocca, with the assistance of the host of thorny Papilionaceæ. The banks of the rivers are also kept up to a very great extent in the same manner, and although high floods often carry away great clumps of the arum, its recuperative power is so great that the line is soon re-established. As the river-banks or shores of the islands become consolidated the seeds of forest trees germinate, and push through the low bushes until they can spread their canopies of foliage over their more humble rivals, and smother them; then we get a bit of the forest where once flowed the river.

When we see such a grand work continually in

progress there is no difficulty in appreciating the fact that the fittest may survive even by chance. While quite prepared to doubt that fortuitous circumstances alone decide as to what shall or shall not remain, we are willing to allow that a great deal depends on accident. In the Essequebo river are hundreds of islands in all stages of growth, each forming a little world of its own, and covered with dense forest or jungle. Those near to the mouth have been mainly the work of the courida and mangrove, of which we shall have something to say in another chapter, but away from the mingling of sea-water all the others have been gained from the flood by the fortunate deposit of seeds and plants floated down by itself. Most of the forest trees provide their seeds with spongy or cellular pods, by which they are carried away to long distances and deposited by accident along the banks of every river and creek. Were it not for the openings produced by the flood hardly a seed would ever get an opportunity of gaining a position for itself, but as it is the water undermines one forest giant and carries seed to replace it by another. It almost seems at first as if the mocca-mocca provides a breakwater for the other plants, but when we see that they have been pushing it farther and farther into the stream for ages, until it has accommodated itself to adverse circumstances, such an idea is impossible. No, this great arum is as selfish as the rest, and cares nothing for those coming behind as long as they do not interfere with him. When they spread their great branches over his foliage, and he can hardly see the sun, his energies are crippled, and he sinks down to a puny dwarf of three or four feet high, or perhaps is killed altogether. Nevertheless we must give him credit for his good work whether he has meant to do it or not; thank him that our riverbanks are comparatively stable, and that some of the islands he has been instrumental in erecting are habitable.

A similar work to that going on in the river is also being performed in the tortuous creeks. Their courses are continually changing through the struggle between the trees and the flood. That the struggle is intense can easily be seen by the naturalist, and even an ordinary observer must recognise the innumerable signs of its presence. As he is paddled round bend after bend during the flood the strained muscles of his negroes' shoulders indicate that they are pulling against a mighty force. Then he comes across fallen trees, floating logs, uprooted palms, great gaps in the wall of foliage, and sometimes large accumulations which check his progress until a passage is cut. Then, also, he sees great clumps of palms (Bactris), beds of marantas, and large masses of tree roots extending out into the water, and can understand that as they increase the channel must become narrower and narrower. That tree which towers so far overhead has an immense cluster of roots which are washed perfectly clean on the creek bank, and even overhang it. If it were not that its branches are wedged and interlaced so closely with its neighbours you would think it dangerous to pass under. It has gone on for years encroaching farther and farther, and now the creek begins to resent the consequent narrowing of its channel. Already it has given the forest giant a warning by excavating a deep hole under its roots, and now only waits that increase of power which will come with the heavy rains to overthrow the stately giant.

That clump of thorny palms, so impenetrable to both man and beast, is also at work vainly attempting to curb the powerful stream. One sucker after another comes up, each a few inches beyond its neighbour, and in a year or two the clump forms a little headland, which drives the stream to deepen its channel on the opposite side. During the long, dry season, while the water was low, the palm made considerable headway, getting a good hold in the mud, until now it bids fair, if not checked, to close the passage altogether. You almost graze it as your bateau is paddled through, and the negroes cry, "Look out for pimpler (thorn), boss!" a warning by no means superfluous, as its long needles might cause serious injury were they drawn across the face.

By and by the flood comes—the channel is not wide enough—the water boils and eddies behind the great root or clump, carrying off great masses of clay and washing the roots clean. We can almost fancy the palm standing up defiantly while the flood is raging to get past. Something must be done, and as the palm will not give way the stream clears a passage behind and turns it into a little island. Perhaps the flood is very high and strong, and the palm clump stands in the midst of a raging torrent, filled with floating logs and uprooted bushes, which are checked by the obstruction and piled one upon another until the whole forms a dam, which raises the water behind to the height of several inches. The stream is running at the rate of six miles an hour, or even more, and as it meets this check it rages behind, foams at the top, runs through at accelerated speed, and presses its tons of water against the barrier.

Suddenly the clump of palms is torn up bodily, the mass floats down-stream, clearing away a hundred other obstructions, and goes down to the sea, to be perhaps thrown ashore a perfect tangle of fibres and broken stems.

When a giant mora is undermined by the flood, and can no longer be supported by its weaker neighbours, it comes down with a great crash, carrying destruction to everything in its way. A score of smaller trees will have their heads torn off or limbs severed, and perhaps a hundred palms marantas, and low bushes, be smashed to pieces. The great branches get broken in the fall, but still the tree rarely rests on the ground as there is such a great heap of *débris* under its head. At the other end the great mass of roots has been driven backwards, ploughing up a groove, which is immediately taken over by the flood and excavated deeper to make up for the partial obstruction of the immense bole. Now comes the grand work of clearing, or stowing away, such an immense trunk, which is not accomplished for several years. First the branches are invaded by hordes of termites, which build their great black nests in the upper forks, and tunnel every part down to the great trunk. As this weakens the props they break off and the trunk settles down more and more until it, rests upon the earth. Perhaps what was once the lower end, near the roots, is actually in the water, and then a barrier is formed against which the current is continually raging. The work of the termites still goes on in that part above water until it crumbles by its own weight, and in a year or so the only portion not decayed and rotten is that which is actually immersed. The flood finds no very great difficulty in dealing with this—the heavy rains fall, great floating masses come down and get jammed under the log, the ends are excavated from their resting-places, pushed this way and that, carried some distance by the floating raft, and finally dropped in some deep place to remain for ages and go under the name of a tacouba.

These tacoubas litter the bottom of every river, but it is only in the smaller creeks, when the water gets low, that they can be seen. Within the tidal influence at low water the swift current is seen running over and among them with the velocity of a rapid. At such times the tacoubas endanger a frail craft by uplifting or boring holes as it comes swiftly down the stream, notwithstanding the Indian's skill with his steering paddle. Like the oak logs found in peat, these tacoubas seem to harden under water, and it is impossible to estimate their age. Like rocks, they are undoubtedly worn away by the continual wash, as may be seen from their rounded surfaces, but otherwise they may be considered almost everlasting. In the old creek beds they are found in the midst of the sand, clay, and pebbles, even then often preserved for an indefinite period, although becoming brittle, and even soft, in a porous soil. From borings made in the alluvion of the coast it has been found that this semi-fossil wood still remains undecomposed at depths of over a hundred feet, being only reduced to its elements when brought to the surface.

As development is only possible through a multitude of generations, the work of the river and creek is of the utmost importance. There must be destruction to make room for new individuals. Without this the same trees would live on for ages, and once the forest was occupied no room could be found for a single seedling. When, therefore, we see that the country is intersected with such a multitude of streams, and that these are continually changing their courses more or less, at each slight alteration giving an opportunity for some new seedling to come to the front, we can understand why species are by no means at a standstill. Instead of stagnation there are continual changes taking place, and it may be safely affirmed that every gully and level spot in the whole country has been at some time or other the bed of a creek. From the battle with the waters the species comes out stronger and more fitted to continue the struggle, notwithstanding that the individual has been overcome.

When we contemplate this destruction from one point of view it seems most distressing. A magnificent giant of the forest towers above our heads to-day, and to-morrow it has fallen and carried destruction to others in its descent. The work of centuries has been undone in a few minutes, and there lies the victim of the ruthless flood. But every year, or twice a year, the warrior has been doing his level best to scatter his progeny over the surface of his enemy, to be floated through hundreds of miles of country and carry on the struggle wherever there is an opportunity. Some of these will no doubt be stronger and better fitted for the strife than their parent and therefore be an honour to the family. The individual in nature lives in his children, and as long as any of his descendants remain we cannot say he is really dead. No, a step in the ladder of evolution has broken, but as the climber will never require to retrace his steps the broken foothold has become quite useless.

VII.

UP IN THE TREES.

NEITHER on the banks of the creek nor in the forest is there room for anything like the smaller herbaceous plants so common in English woods. Anemones and bluebells flower before the trees put on their summer vesture, when plenty of light is obtainable; here the forest giants are never leafless. At no time can the shady wood be compared with the gloomy forest, and therefore the presence of so many pretty flowers under the trees, even in summer, is easily explainable. In Guiana, except two or three saprophytes—Voyrias and a tiny orchid, reduced to nothing but weak and almost colourless stems covered with minute scales instead of leaves, with one or more flowers—there are absolutely no plants on the ground. Even the Voyrias require a little more light than they can generally find, and are therefore wanting under the densest shade.

To see the representatives of the pretty wood flora of temperate climes we must look overhead, and even then find nothing unless we go to the banks of the creek, the edge of the forest, or the sand-reef. In the recesses there is absolutely nought but bare trunks and leafless bush-ropes. Even the epiphytes want "light, more light," and without it cannot exist. Where, however, enough of this precious influence is obtainable they crowd every branch and twig almost to the ground, and carry the struggle for life right up to the tree-tops. Here is a little world in itself—a world only represented in temperate climes by a few mosses and lichens, with here and there a fern. Monster arums twelve feet in diameter occupy the great forks, and throw down long, cord-like aerial roots from their nest-like rosettes of great arrow or heart-shaped leaves. Looking up as we push these cords aside, the plants are barely discernible, on account of the crowd of other epiphytes which surround them. Screens of creepers, with their festoons of handsome flowers, masses of the mistletoe-like Rhipsalis Cassytha, pendulous branches of grass-like ferns, and a thousand epiphytes on every branch, obscure the view and make it hard to say from whence a particular aerial root is derived. Some branches

SILK COTTON-TREE CROWDED WITH EPIPHYTES.

are occupied by dense rows of Tillandsias, which, as it were, push everything else aside and take possession of the upper surface. They can only grow upright, as their vase-like circles of leaves form reservoirs of water against the time when little or no rain falls. Other plants have had to store up moisture in many different ways, but this is probably the simplest, and gives less trouble than the building up of an assemblage of cells. In the mountains similar reservoirs are utilised by the beautiful Utricularia Humboldtii, as if they were pools—a striking example of the many shifts and expedients of plant life to prevent a single ray of sunlight being wasted.

These are by no means all the epiphytes. Hardly a twig is free from them unless the gloom be too great. Fortunately most of them have attained the power of living on very little light, as certain others can exist and thrive in places where we might expect them to be burnt up. Yet, although these more delicate epiphytes never see the sun, they differ much as to the amount of light they require. On the boles of the trees, often below the level of even ordinary floods, grow patches of filmy ferns, the leaves in some species overlapping each other and entirely covering the bark for several feet. A little higher comes the Polypodium pilosclloides, and above that the Peperomia nummularifolia, both of which spread their thread-like stems in every direction, decked with pretty oval or circular leaves, covering even the tiniest branches. Among them hepaticæ and mosses spring up, and on these grow little orchids with almost microscopic flowers, all combining to give a festive appearance to the whole. These tiny beauties are, however, rarely seen from the creek—in front there is only a bank of foliage which shuts off everything behind. Pushing our bateau through the screen, however, and looking up, one of the prettiest sights in the world is revealed. In the half-light, which is so grateful to the eyes as well as the feelings, we see the main stem rise like a pillar, clothed in green, and decked perhaps here and there with the beautiful flowers of that pretty orchid, Zygopetalon rostratum. From this rise at intervals numerous garland-covered branches, on which are seated the rosettes and clumps of such plants as Pleurothallis, Stelis, and Specklinia, some of them unequalled in beauty, although excelled in size by the larger members of their family.

Reluctantly leaving this fairy bower, we again come out into the creek and look up to a great branch which extends almost horizontally a few yards above our heads. On its upper edge the inevitable line of Tillandsias has taken the first place, pushing every other plant aside to do the best it can. However, there are one or two plants not disposed of quite so easily. The Rhipsalis pachytera, that curious ribbon-like cactus, is strong enough to insinuate himself into the midst of the enemy, and make it stand the strain of his long archings and pendulous extensions. It almost seems as if this were done deliberately—the Tillandsia tried to push off the cactus, the latter secured himself by holding on to the hustler, with the result that the one has to hold the other up, or both will fall into the creek and be drowned. The beautiful golden fern, Chrysodium vulgare, has also succeeded in pushing its scaly rhizomes among the roots of the Tillandsia, and hangs its fronds in every direction round the branch. By means of its creeping habit it is able to grow in any direction, and is therefore very common, not only on the branches of trees, but on their stems as well. Among the plants which have been driven to the edge are some of the most beautiful orchids. Doubtless they at one time grew upright, but from the continual pressure of circumstances they have become perfectly fitted to their environment. We may fancy them in some past age asking the question, "Shall we develop ourselves to fight the enemy, or get out of the way?" As gentlemen they chose the latter course, with the result that their progress has been towards greater perfection in form and colour, to the exclusion of everything hard and disagreeable. Their enemy, on the contrary, has gone on with his bullying, until now he is often clothed with spines, and perhaps as disagreeable, to our sense of beauty, as the orchid is harmonious.

There, on the side of a branch, is a fine clump of Brassia Lawrenceana, its spikes of yellow flowers dotted with crimson arching gracefully over, and apparently quite content to grow sideways. Like many other orchids of the same habit, however, the Brassias are not only able to grow upright as well as inclined, but also to bear crowding. This is not the case with its neighbour, the Stanhopea eburnea, which, sitting a little below, pushes its two or three handsome, waxy-white flowers from below the pseudo-bulbs outwards and downwards. Although the plant itself is almost upright, its short flower-stems could not reach the light without an opening below, nor could they do anything if crowded on the upper edge of its support. Another stage in the progress of accommodation is shown close by in the Gongora atropurpurea.

This plant also is able to grow upright, but its lax flower-spike is pendulous, allowing the curious, grasshopper-like flowers to hang two feet or more below the branch. But even yet we have not come to the end of these contrivances, for the Scuticaria can actually grow underneath the limb and hang its flexible, cord-like leaves straight downwards—in fact, although the plant may grow on the side, it is utterly impossible for it to stand upright. As may be supposed, it has few competitors for this position, and can be considered as having reached the highest point on this particular line. A similar development is exemplified in an epiphytal class of ferns (Acrostichum). Some species have short, thick, leathery leaves, and grow either upright or leaning; another division has the leaves thinner, more or less pendant, but still short; while at least one species hangs down like a bundle of long, flexible ribbons.

We have spoken of the reservoir of water contained in the Bromelias, and must now deal with the contrivances of other plants for preserving life during a long drought. For, although the rainfall is so heavy, nevertheless there is generally an intermission of at least two consecutive months in every year. Then the epiphytes would wither and die if some provision were not made against such a contingency. Even the little pool of the Bromelia does not last many weeks; but not content with that, the far-seeing plant has taken care to reduce the evaporation from its leaves to a minimum and made them hard and horny. In epiphytal ferns moisture is stored in thick, leathery, or scaly leaves, and in rhizomes of a similar nature. In the orchids, however, we have the pseudo-bulb—a unique contrivance by which moisture is retained through the longest drought. When the leaves are thin they are often dropped, and the plant becomes nothing more than a bundle of green pseudo-bulbs attached to its support by a few almost dry aerial roots. In this dormant condition it rests quietly, and uses the store of moisture simply for the purpose of keeping itself alive. Other species have smaller pseudo-bulbs, and thicker leaves, which are not deciduous—these rest somewhat, but never to the same extent as the first. Finally, we have species without any means of storage except their thick, leathery leaves, and these appear to endure drought with less injury than the others.

That these epiphytes—air-plants, as they are sometimes called—are able to live in the way they do, shows a marvellous power of development in some past age. Some of them have got to the top of the tree, actually as well as metaphorically, and we can hardly conceive of further advance in that direction. Here and there we find indications of the lines they have followed, especially among the orchids, which, taken altogether, are undoubtedly ahead of all the rest. Thus the genus Catasetum contains some species which live on the sand-reef, others growing on low trees in the swamp, and one which has found a congenial home among the leaf-stalks of the eta palm. In the pure white sand there is little more nourishment to be obtained than on the trees—in both places the plant can obtain virtually nothing but air and moisture. Going down to the pipe-clay savannah, is again but a short step as the soil here is hardly more fertile than the sand. Again, on the mountains, orchids flourish indifferently on rocks or low trees, and it can easily be understood that a tree growing in a crevice may carry with it as it rises upward the plant which has made itself fast to its young stem or branches.

However wonderful the habitat of the orchid may be, it can not be compared with its other developments. As we have seen, the leap into the trees was but a short one, for which it had long been prepared; but what shall we say to the construction of a home for ants, so that its tender aerial roots may be protected from cockroaches and other pests? This many species have accomplished, and now do it so thoroughly as to derive considerable benefit from the contrivance. Perhaps the most perfect of these homes are those provided by Schomburgkia and Diacrium bicornutum. In them we have a hollow pseudo-bulb, into which the ants either find a doorway ready made, or are offered inducements to make one for themselves. The result is a perfectly dry, hollow chamber, on splitting which the tiers of cells and galleries are seen ranged from top to bottom. Another and quite distinct harbour for an ant garrison is that of the Coryanthes. There are several species, all of which appear to grow in the same manner, attached to bush-ropes rather than perched on limbs of trees. This is so obviously suited to their peculiar manner of growth that it is quite conceivable the plants may have been first carried up from the sand by the tightening of the stem of a creeper as a growing tree carried its mass of foliage higher and higher. Otherwise, we might fancy they pushed off from a branch and being caught as they fell to make themselves at home under new conditions. However this may have been, the fact remains that the plant is

AN ORCHID ASSOCIATED WITH OTHER EPIPHYTES.

now entirely suited to its habitat, and flourishes to perfection in places where few other orchids succeed in establishing themselves. Instead of a hollow pseudo-bulb, the Coryanthes provides an oval mass of fibrous roots, as distinct from those so well known in the other orchids as their object is different. In the Coryanthes the ants establish themselves, filling up the interstices to make a waterproof nest, whence they are ready to issue on the least alarm of an enemy. Being carnivorous they can do the plant no harm, but, on the contrary, are so useful that without them it suffers greatly from cockroaches and other pests. This is easily proved by specimens brought to our gardens, where, on account of the collectors having removed their useful tenants by soaking them in water, they are particularly subject to the attacks of insects and rarely thrive for any length of time. The nearly-allied Gongoras also make a less perfect provision for ants, and the great Oncidium altissimum often has such large communities that the collector finds it very difficult to dislodge the plant from its perch without getting severely bitten. Here we see a gradual transition from long aerial roots running in every direction, and a massing together of an intricate maze of fibres.

The long aerial roots, so common in orchids, are mainly concerned in securing a firm hold to their support, and when a plant is removed to a new locality all its energies are devoted to this end. The orchid seems to know how important this is, and uses great discrimination in utilising its knowledge. It is apparently capable of choosing where and how the attachment shall be made, as although there is no doubt great similarity in the modes adopted by individuals of the same species, still close observation shows that they are not identical. The first precaution taken is to secure itself immediately under the bunch of pseudo-bulbs, and in this it takes so much care that we can hardly conceive of such excellent results coming from anything less than forethought. The plant may be perhaps wired to a block or placed in a basket. In the first case, notwithstanding the fact that it is perfectly safe from falling, it provides against contingencies by firmly attaching itself; while in the other it is more inclined to throw out a number of arms as it were, and twine them round the bars. If the growths are very long they of course exert a powerful leverage, and the attachment requires to be all the stronger. Then the aerial roots wind round the branch and extend upwards and downwards for yards, clinging so tightly that they mould themselves as it were to every little channel and roughness of the bark. At the end is a green point about half an inch in length which is peculiarly sensitive and such a luscious morsel to cockroaches that the plant, as we have already seen, has in many cases provided a garrison to protect it. When the plant has fastened itself securely, more aerial roots are produced which go wandering in every direction, sometimes running among the creepers, the roots of ferns, over the mosses, or hanging downward without any attachment whatever. If there is another branch near the one on which they sit they grow towards it, and thus secure a second hold in case of accident to the first. The benefit to be derived from this can easily be seen when a branch dies and begins to decay. Then the sensitive aerial roots seem to appreciate what is about to happen and loose their hold. The sensitive points turn away in apparent disgust, the whole plant shrinks from what is poisonous to it, and will rather allow itself to fall into the midst of the thicket below than run the risk of further contamination. Then comes in the advantage of having a second perch which has perhaps remained healthy. As the plant looses itself from the one, it is drawn towards the other, and, perhaps slipping off, hangs downward as if held by a number of strings. In such a position it has great difficulty in recovering itself, but in time again raises its head and is apparently as flourishing as before. Although the aerial roots are mainly concerned with securing the orchid on its perch, they also appear to help the leaves, by absorbing moisture, and even in some cases to assimilate the vegetable infusions which continually trickle down the tree-trunk and branches. This liquid manure, however, does not seem to be absolutely necessary, although it may probably be of advantage when obtainable.

We have seen how different species accommodate themselves to their positions and produce nodding or drooping flower-spikes to suit. This is just as strikingly exemplified in different individuals. When the flower-spike is free from obstruction it grows outward or downward as the case may be, without altering its course, but when it is crowded by other plants or its own pseudo-bulbs the case is different. Then, the young spike grows upright as far as is necessary, and then gracefully arches over until the flowers can hang free. It follows therefore that the peduncle is longer or shorter according to circumstances.

Again, such plants as the Stanhopea appear to have the power of selection to a great extent, possibly even so far as to choose a particular spot on the outside of its mass of bulbs for the flower stems. Under cultivation in open baskets they commonly push them straight downwards through the bars, which is obviously almost impossible under natural conditions. It also seems as if the plant calculated the shape and size of the opening before producing its flower-spike, as amidst such a crowd it might easily happen that the space would be too small for the proper development of its flowers. Of course it makes a mistake sometimes, with the result that the bud rots or the great blossoms get smothered, but this occurs very rarely.

We have lately been investigating the life history of the species of Coryanthes and especially C. macrantha. This genus, as we have already mentioned, has developed an oval mass of roots to accommodate a garrison of ants. But it is not in this, or even in the selection of a habitat, that its powers are most highly developed. No, when we come to the flowers we have perhaps the most wonderful arrangement in the whole vegetable kingdom. When we see the pair of great flowers at the end of their pendant stalk we wonder how such monsters can be derived from so small a pseudo-bulb. Sometimes the mass of orchid roots is occupied by a host of rivals—in a specimen we have there are two distinct species of Coryanthes, a Bromelia, several Anthuriums, and a young shrub, all of which are growing on a ball hardly a foot in diameter. It follows therefore that in such a crowd the plant has some difficulty in finding a place for the development of its flower-stem. Nevertheless it does this and does it well. From between the crowded bulbs comes up the bud, and you may fancy from its direction that it is about to extend upwards and get above the crowd. But it cannot go quite so far as that. As it reaches the apex of the bulb-cluster it almost seems to take a survey of its position to find out the best and roomiest place for its handsome twins to occupy. Having decided, it arches gracefully over and grows straight downward, enlarging its flower-buds as it goes along. By and by you see that there will be two flowers and as they now stand, one is sessile above the other and obviously too close to have room to expand. Now the peculiar Chinese-foot-like shape begins to appear, but still the upper bud remains without a stalk. Presently, however, the whole stem becomes enlarged, the point to which the lowest flower-bud is attached turns outward a little, and the pedicel gradually changes its direction from perpendicular to an angle of forty-five degrees. Meanwhile the second bud has developed its own stalk which takes the opposite direction, thus forming two lines of a triangle with buds at the bases. Now the two unopened flowers stand well away from each other, and the plant seems measuring the proper distance which will allow of their opening without standing in the way of each other. This object having been attained almost to a line's-breadth and the buds of their full size, you come down one fine morning to be astonished at the transformation. The tissues which swathed the foot have bent back and revealed a pretty fairy cup into which drips at intervals from two horn-like processes above, tiny drops of nectar.

Darwin has described a few of the peculiar contrivances by which orchids are fertilised, and some of his deductions from the conformation of their parts drawn without actual observation of the insects at work show the marvellous insight of a great mind. But how he would have delighted in seeing the insects at work before his very eyes! To read of the fertilisation of the Coryanthes or Catasetum is one thing, to observe it is another.

When you come down in the morning and see a host of beautiful metallic green and gold bees hovering round the orchids, you know at once that the Coryanthes speciosa which you admired in bud yesterday is now open. How these bees, which never appear at other times, have made the discovery it is impossible to tell. There is a slight perfume diffused round the flowers, but it is by no means pungent, nor does it extend to any distance as far as our sense of smell can distinguish. Yet the bees are here buzzing around, creeping under the cap-like appendage of the flower, and then flying off or dropping into the little pool below. Looking inside we see one of them floundering in the shallow liquid, its wings bedraggled, in a mess and unable to extricate itself. It struggles to climb the slippery sides of the cup, but all its attempts being useless, it goes swimming round and round until almost exhausted. Presently, however, it spies a gleam of light coming through a mouth-like slit where the apex of the column approaches the cup but does not actually touch it. The approach to this slopes upward, and is very slippery, but the insect inserts its fore-legs into two little gaps apparently provided for this purpose and drags its head into the opening. This, however, is too narrow to permit of its exit without trouble, but at the same time, being springy, offers no real obstruction. You see the insect straining to get through, its head moving now this side, now that, as the right or left leg is brought into play. Presently with a jerk the body comes through, and the almost exhausted bee crawls slowly away. But what is that sticking between its shoulders, a conspicuous double mass of yellow on the green? This is the pollen which has had the opportunity of fixing itself in that position during the bee's struggle to get out, and the foolish insect stumbling into another flower on the same spike, or perhaps on another plant, unwittingly fertilises it, and thus completes the work for which this elaborate apparatus was contrived.

The species of Catasetum are almost as interesting, but here it is a large humble-bee, black with yellow bars across the abdomen, which is the fertilising agent. Like the other, it is almost a stranger to our gardens, and appears in a similar manner as soon as the flowers open, even when they are hidden under a dense canopy of foliage. Flying into the flower the bee's proboscis comes in contact with a sensitive, antenna-like appendage connected with the little box in which the pollen is contained. Out jump the pair of pollen masses as if let fly by a spring; they adhere to the back of the insect by their peculiar sticky disk, and are thus carried from flower to flower to fertilise others than the one in which they were developed.

Of course we cannot stand up and watch these operations when paddling up a creek; we must bring the plants into our gardens on the coast to do that. This is unfortunate in many cases, as the insects necessary for particular species only live in or near the forest. Here and there we discover that a white flower is fertilised by a moth, or suspect that certain others benefit from the visits of butterflies; but beyond the two genera we have mentioned no others have come under close observation. Possibly there may be operations going on in other genera of which we know little or nothing, almost as interesting as those in Coryanthes and Catasetum; but we can only guess the purpose of their contrivances, which are so numerous and puzzling.

VIII.

IN THE SWAMP.

BEHIND the fringe of couridas that guards the coast of Guiana lies an extensive swamp, which on being drained becomes one of the most fertile soils in the world for the sugar-cane. At present, however, we are not concerned with its suitability for plantations, but rather have to deal with it as it exists apart from man's interference. Although at first sight it is nothing but a dreary waste of sedges rising from the oozy morass, we shall find on close examination that there is a continual struggle going on here as in the forest. We have seen how intense this is among the trees, and it might be supposed that the weaker inhabitants of the swamp would be unable to contend with each other like their powerful cousins. When it is considered, however, that there is not only a straining after light, but also protective contrivances against flood, drought, and fire, to be developed, we see that the dense carpet of vegetation can by no means remain idle.

If it were not for the creeks which meander through the swamp any attempt to penetrate would be hopeless except at the end of a long dry season, but by means of these natural drains we can enjoy the sight of nature's handiwork without much difficulty. Perhaps we may approach it through a mangrove swamp or a line of tall trees overhanging the lower reaches of the creek. Here, as in the forest proper, the dark waters are shaded, but the arcade is not so dark nor so clear from obstructions. Not only are the trees interesting from their variety, but the wealth of climbers, epiphytes, and parasites make every turning appear more beautiful and interesting than the last. The current is very strong, and at each bend the stalwart negro boatmen have to strain at their paddles in a way that conclusively proves its power. Now we are at a standstill for a minute or two, as the stream appears to resent our intrusion and comes round a little headland with amazing rapidity. Sometimes we are even driven back among the thorny bushes, and have to look out that our faces be not scratched. On we go, however, although but slowly, until the current becomes weaker, and we emerge from shade into bright sunlight. At first our eyes are dazzled by the contrast, and we can hardly look up, but as they become more accustomed to the glare we can see over the wide meadow-like expanse which stretches for miles on either hand.

Here at least it may be said that there is light enough for everything to have its share. But when we observe how dense are the sedges, and that even the creek is almost choked with vegetation, this opinion is expressed with some hesitation. Let us look a little closer, and what do we find? Why, the struggle for life is even more intense than in the forest! There the trees extend themselves over each other and form great canopies; here the flood will not admit of such a thing. Anything that could be taken hold of by the water would quickly be uprooted and carried off; it has followed, therefore, that the form of leaf which offers least resistance has been chosen. This is undoubtedly the long and tapering foliage of grasses and sedges. These admit of a rise and fall almost to any extent, and being flexible accommodate themselves also to a pretty strong current. How densely they are packed! We might fancy that the sunlight would penetrate right down to the ground; but drag up one of the sedges, and its base is seen to be bleached as if it grew in the dark. And that this is the case there can be no doubt. If it were possible for us to walk among them without pushing a leaf aside we should certainly find the shade at least as dense as under the trees.

Such a crowded assemblage of vertical leaves must necessarily gather an enormous amount of light, and that this is so can easily be understood from the marvellous energy displayed by these grasses and sedges. With an unlimited supply of water, plenty of heat, and as much light as they can wrest from their neighbours, we can almost see them grow. An English meadow, when almost ready for the scythe, is the scene of a similar struggle, but how insignificant in comparison with this. There the highest grasses are hardly more than three feet above the ground, here they rise to ten feet, or even more. In crossing a meadow the grasses are easily borne down before you; but in the Guiana swamp, if the ground were at all dry, you would have to press forward with your head and arms, use considerable force to make a passage, and even then be unable to find your way. For the thicket almost closes again behind as you press along, and the deep rut shuts out everything but the sky above.

We once had an experience of this kind in the

dry season when the bottom of the savannah was only a little oozy. From our bateau on the creek we had seen the characteristic ribbon-like leaves of Catasetum longifolium—one of our most beautiful orchids—hanging down beneath the canopy of an eta palm about fifty yards away. Nothing would satisfy us unless we at least made an attempt to gather it, and notwithstanding hints of monstrous boas and venomous insects we stepped ashore. The sedges were comparatively light, but even then it was hard to force a passage. First we tried leaning forward with our face protected by our hat, at the same time dividing the sedges with our arms, but some scratches from razor grasses soon showed that this method was hardly practicable. Then we turned our head and backed for what we estimated to be the right distance, but could see nothing of the palm. We had, as we thought, moved in the right direction, yet, in the absence of anything like a landmark, we stumbled along, pushing to this side and that, and ultimately had to give up the search. And then, the feeling of loneliness and isolation was indescribable. True, we could call to our friends in the bateau, but could see nothing of them any more than they could of us. Fortunately, our path was easily retraced on account of the way the sedges had been thrust aside, and we got back to see the beautiful streamers of the orchid still hanging in a most tantalising manner within apparently such easy reach.

During the rainy season, however, it would be impracticable even to make such an attempt. Then the apparently green field is covered with water varying in depth from one to three feet. Sometimes one of the negro boatmen is sent into this morass to bring in a dead bird that has been shot, but rarely does he find it unless it is so close that both bird and man can be seen from the creek. Down he flounders into the water, now rising on a tuft almost to the surface, and then slipping to his middle in water and ooze, or perhaps sprawling flat and cutting his hands with the razor grass in his attempts to save himself. As he cannot see where he is going there is a continual hallooing of "right" or "left" as he blunders along, and sometimes abuse for not knowing over well the meaning of these words. The sportsmen in the boat can see the muscovy duck very well, but the man in the water will do much better than either of them could if he gets it.

The places we have been attempting to describe are the most accessible portions of the swamp— what shall we say of the great tracts where the mighty razor grass comes to the front and turns out everything else? Here there is no question as to its impenetrability in either dry or wet season. The sword-shaped leaves of the monster are armed on edge and keel with the most beautiful saws that can be imagined, which are so sharp as to make deep gashes in your face or hands almost before you are aware you have touched them. Only the scaly alligator can make a way through the tufts of this monster, and even he is held in check by the density of its growth. In some places it covers miles of the swamp to the exclusion of everything else, its great brown panicles at certain seasons rising six or eight feet above the surface and giving a ruddy tint to the otherwise green expanse. Not only is it well protected against floods, herbivorous animals and rival plants, but is able to endure fire as well.

When the dry season comes and the morass is thoroughly drained, the razor grass begins to look parched, and if no rain falls for three or four months its outer leaves shrivel and lose their colour. At such times a spark from the fisherman's camp-fire, or his pipe, comes in contact with some of these and the savannah soon bursts into flames. In the day you can see clouds of smoke running along, and hear a continual hiss and crackle, with now and again reports like pistol-shots, but, on account of the intense sunlight, see but little of the flames which are running along the ground and demolishing one great tuft after another. At night the scene is magnificent. For miles the ground glows like a furnace and the flames shoot up now and then in pyramids and great sheets as they spread around in a circle until checked by the belt of eta palms and forest trees in the far distance. When the fire has spent itself, which is not for several days, the once beautiful green expanse is an ugly black field from which even a zephyr raises a cloud of charred particles. You can now walk upon it if you do not mind the choking dust which rises at every footstep. All that remains of the king of the swamp—the mighty razor grass—are blackened tufts which cover the ground at distances of about two feet from each other with narrow channels between. If you look closely you will find that the ground itself has been burnt and that it has sunk for about a foot. Your feet go down into it at every step, in some places almost up to the knees, and you get so covered with the flakes as to appear almost like a chimney-sweeper. The bottom of the swamp is covered with an oozy kind of peat called pegass, and it was this which burnt with the furnace-like glow at night. It is this also that tinges all the water of the creeks, making them of the well-known coffee colour which has given rise to the name of Rio Negro, so often applied to South American rivers. Unlike peat it is not fibrous, but apparently made up of layers, which in dry seasons can be separated one from another, and proved to consist of thin strata of leaves. It is somewhat elastic under foot when dry, but sinks at every step with a crunching noise on account of its brittleness. When drained and put under cultivation, pegass lands gradually sink until their level is lowered about two feet, as the vegetable matter being exposed undergoes thorough decomposition in the full sunlight. Like the reduction of fallen trees in the forest to a rich humus, the breaking down of the cast-off leaves in the swamp is an entirely different operation to that which goes on under other conditions, and is well worth studying. The main factors concerned are moisture and the absence of direct sunlight, perhaps combined with the antiseptic properties of tannin which is found more or less in the bark and roots of all the trees and plants. However, we must return to the great razor grass, which might now be thought gone past recovery. Nothing remains but its blackened tufts and you think it a good riddance to such a pest. Perhaps the thousands of weaker plants would think so too if they had the power of reasoning. At any rate, when the rains fall and the flood again covers the blackened surface, myriads of seeds are scattered everywhere, to sprout and give it again that beautiful meadow-like surface. But the monster is scotched, not killed, and soon every tuft is throwing out new shoots to come into competition with the late arrivals. Then ensues a great struggle, the end of which can easily be predicted. The beautiful saws lacerate everything in their way, smother their pretty rivals, and the monster is soon again master of the field and monarch of all he surveys.

For some reason or other the great razor grass does not succeed in every part of the savannah. Either the soil is too poor or the water too shallow to suit his greedy appetite, and, therefore, he is magnanimous enough to let other plants occupy such places. But even here the struggle for life still goes on, and is perhaps all the more interesting from the varied characters of the combatants. Here the sedges are more delicate, mixed with pretty grasses and variegated by flowering shrubs. Palicouria crocea with its fiery bracts; the pale yellow Jussiæa nervosa; Rhynchantheras with great purple flowers; and Hydrolea spinosa looking like a great thorny borage, combine with ferns (Blechnum serrulatum), aroids, and heliconias, to make up quite a show. Perhaps this part of the savannah may be a great amphitheatre almost surrounded by the forest, which rises as a sloping embankment at the edge, only broken where the creek enters on either side, and even there without this being apparent at any distance. Its most striking characteristic is the fringe of tall eta palms (Mauritia flexuosa), which stands forth in front of the dense wall of foliage and keeps guard as it were against the incursions of the flood. Like the courida on the sea-shore the eta palm has developed to an extent beyond other species, a power of enduring the flood, so that it can exist and thrive under circumstances that would be fatal to other forest trees. Even from its first stage as a young plant it is obviously well fitted to its environment.

Its specific name, *flexuosa*, describes the character of its fan-shaped leaves, which are particularly flexible when rising in the midst of the water before the stem has been developed. That this is a wise precaution against the flood can easily be understood, as it offers little or no resistance to the strongest current. Like so many other plants, the eta raises itself on a great tuft or mound of roots, with the result that when thousands of them extend along a line an imperfect dam is produced, which checks the flow of water to a considerable extent. Another result is that the level of the ground—if we may call that level which is made up of mounds with narrow channels between—is raised several feet.

Leaving the eta swamp we come upon one of the prettiest scenes in this part of the country. The creek meanders through a park-like expanse, with wavy lines of bush alternately approaching and receding from its banks, making great bays which add much to the beauty of the picture. The soil is a barren white clay quite free from pegass, and except for the tufts of fine wiry grasses and sedges, almost like a field of snow. As in other places the ground is very uneven, great clods alternating with narrow channels, filled with milky water, into which our feet continually slide as we try to step from tuft to tuft. As may be supposed, the great monsters of the swamp are starved out of these little paradises and the struggle is almost entirely one for food, the survivors being those which require but little. As a natural consequence foliage is conspicuously meagre and flowers come to the front. At certain seasons this savannah is quite gay—it is a veritable flower garden. Round the borders grow miniature pine-like forests of Lycopodiums, in the midst of which the beautiful blue bells of the Lisianthus hang and relieve the otherwise uniform but pretty mass of foliage. All over the savannah the phlox-like flowers of Sipanea pratensis give a blush of pink to the whole expanse, but not so as to exclude a number of others. In one place every tuft is occupied by a plant of Cleistes rosea, that pretty ground orchid with flowers as brilliant as a Cattleya, although not so large. Then there are Burmannias, a host of Utricularias growing from pretty rosettes, or accompanied by grassy leaves; and the daintiest gem of all, the ruddy Drosera, here as everywhere else entrapping the myriads of microscopic gnats which frequent such places.

Leaving this savannah, on which we have been able to walk and admire the host of pretty flowers that deck the surface, we come back to the creek, which has by no means received the attention it deserves. Its most striking feature, besides its dark waters, is the almost total absence of current in the upper reaches. Whereas, near the mouth, the stream is narrowed by the intrusion of a thousand clumps of palms and marantas, as well as great boles of trees, here it widens out into still lagoons almost choked by the masses of vegetation growing actually in the water. Above all the rest the most conspicuous are the water-lilies, the long flexible leaves and flower-stems of which make double work for the paddlers as they have to be pushed aside at every stroke. In the day their flowers are all closed, but at night they lie open on the surface of the water and distil their powerful fragrance into the damp air until it becomes almost oppressive. To see them on a moonlight night is worth a long journey, as they reflect the rays as if they were fallen stars, especially when their petals are laden with dewdrops. These plants above all others seem as if specially developed to suit the swamp. However shallow or deep the waters may be the leaves always float on the surface. The petiole may be a dozen feet in length or only a single inch when the creek is dry. If a flood rises quickly before it can elongate itself there is no danger, as the flexible stems and leaves can move backwards and forwards without affording the least opportunity for the waters to take hold and drag it from its anchorage.

Between the great leaves of the water-lilies are patches of Utricularias, their beautiful flowers standing up out of the water like yellow and purple violets. Their lace-like foliage spreads

INUNDATED FOREST.

below the surface, its green tracery contrasting with the ruddy-brown fluid in which it is merged. Then we have the Cabomba, with pretty floating shields and finely divided foliage of a circular outline below, which reminds us of a patch of miniature buttercups. Unlike the water-lily and its congeners, these plants are particularly delicate and fragile, yet they flourish to perfection, notwithstanding that their stems are torn to pieces by every flood. Every tiny piece is the parent of a host and the raging waters only serve to disseminate it in all directions, to again come to the front when the creek resumes its wonted stillness.

The swamp is not entirely devoid of trees. Besides the eta palm, the Genipa, the fruit of which is used by the Indians to make the blueblack markings on their skins, and the Tabebuia a species of Bignoniæ, are common. Then there is a host of other species all more or less suited to the swamp and never found in the forest proper. At first sight they appear sickly, and might be taken to indicate a passive endurance of unsuitable conditions. Unlike the giants of the forest they have no dense canopy of foliage far above the ground, but are beset with gnarled branches almost down to the level of the water. Then, their twigs are open and their leaves comparatively few and far between, so that they hardly overshadow the undergrowth of sedges, only preventing it from becoming so dense as in the open savannah. Instead of these characters being signs of degradation however, close observation appears to indicate that, like every other living thing in their neighbourhood, they have been developed to suit special conditions and circumstances. It is obvious that a large and heavy crown would tend to bring down the tree with its weight when loosened by the flood. Therefore, like most of the shrubs under the same circumstances they spread their roots over the pegass and take care not to provide a mass of foliage which would obstruct the flow of the water when it is high, or tend to overthrow them by becoming top-heavy. The effect of all this is beautifully illustrated in the Cabomba; as long as it grows under water, only feathery leaves are produced, but when it reaches the surface it covers the water with its pretty shields. Something similar takes place with the tree; in the swamp its branches are numerous and open, while in the forest it produces a thick canopy of foliage at the top of a naked trunk.

That the struggle for life in the swamp is more intense than that in the forest can easily be understood when such varied aspects are considered. The fight with others and the straining after light are similar in both cases, but in the former we have also the endurance of water, provision against the flood, the ability to flourish where there is practically no food, and, most wonderful of all, the power of the razor grass to endure and recover from its cremation.

IX.

ON THE SAND-REEF AND MOUNTAIN.

WE have seen that with the continual destruction going on in the forest there is ample scope for development. On the sand-reef or mourie, however, this is not so patent, and we should therefore expect to find a more primitive flora in such places. Again, when we get to the mountain region of the interior, the plants will probably be of still more archaic types. Standing as these do high above the point reached by the flood, they have one element the less to contend with, and are therefore all the better enabled to cope with the disabilities peculiar to their position.

The mourie was, as we have said, the sea-shore of some past age, before the alluvion, on which the plantations now stand, was in existence. It consists of reefs rising to a height of about a hundred feet, with gullies and slopes, sometimes narrowed to a ridge, at others broadening out

ON THE LAND-REEF.

into plains of several miles in extent. The vegetation is quite distinct from that of the forest, not only in form but also in kind, consisting of clumps of thick bushes with spaces between, or a thin forest somewhat resembling that of temperate climates. Between these clumps the white sand is often quite bare, and reflects the intense sunlight to such a degree as to be quite painful. It is often so hot that the bare-footed Indian has to cut pieces of bark and make sandals before attempting any long journey across it. Sometimes we come upon open spaces of a moderate size, where a thin wiry carpet of under-shrubs manages to exist, mixed with a few annuals in the rainy seasons. These little shrubs consist of several species of Papilionaceæ, Cinchonaceæ, and Melastomaceæ, and the annuals of a Polygala or two, and a few grasses. In such places also the pine-apple is common, not perhaps truly wild, but probably the offspring of the head of a fruit thrown down there in some past time. Many of the plants on these places are very pretty, with a heath-like habit, and when in flower give the mourie quite a gay appearance. All over the sand are myriads of ant-tracks which remind us of the runs near a rabbit warren in England.

It is not however the open places that the

naturalist finds most interesting, but rather the borders of the clumps of bushes, where, lightly shaded, an assemblage of very interesting plants are growing. Perhaps the most striking to a newcomer are the species of Pæpalanthus, which grow in tufts, their rosettes of whitish hairy leaves, in some kinds hardly bigger than the thumb-nail, reminding him of the pretty cushion-pink so well known in English gardens. Some species, very common on the dry savannahs of the interior, are much larger, with tufts as big as the top of a child's head, their whitish bristly leaves closely resembling the unkempt hair of some ragged urchin. Another curiosity is the Schizæa dichotoma, a fern, the leaves of which are more like those of a wiry sedge than others of its family. But by far the most handsome plants in such places are the Cyrtopodiums, great orchids with long pseudo-bulbs, rising two or three feet above the sand, overtopping which stand panicles of splendid golden flowers. Like some of the epiphytes these harbour nests of ants among their roots, and although it might easily be said that they have come there by accident, we are of opinion that such is not altogether the case. Here also grows the Catasetum discolor, its curious hood-shaped flowers and general habit showing signs that it is more archaic in its type than the epiphytal species of the same genus. All about these places lichens and mosses are common, on the sand as well as on the bushes, the cushions of Polytrichum and Octoblepharum often covering large surfaces.

But, perhaps the most interesting plants on the mourie are those which are also found in the forest under different conditions. The Clusias and figs are living quite a harmless life—they no longer perch on the trees to strangle them, but use their best endeavours to get over the diffi-. culties of such a hot and barren soil. There is so much work to be done in the way of groping down into the cooler depths for moisture and the weak infusion of vegetable matter, that they have hardly time to attend to other business.

Instead of climbing upwards to fight for a share of the sunlight they have rather to harden themselves against it, as here it is so plentiful, so glaring, and so very hot. The forest giants manage to exist here and there on the slopes, but they can only do so by modifying themselves so much, that woodcutters distinguish between the timber of such places and that of the swamp as being harder and much more durable. Then, we have a wealth of scrambling vines, which, instead of climbing to the tree-tops, run along the sand and show the different conditions by their paucity of leaves, hardness of stems, and brilliancy of colouring. The shrubs too look poor and meagre, and are more subject to the attacks of the bloodsucking loranths, although even these pests cannot succeed in forming such large and dense masses as in other places.

The vegetation here may be looked upon from two points of view; first, that which represents it as an approach to the ancestry of the trees and epiphytes of the forest; and second, as a class which has never succeeded in migrating beyond its peculiar environment. As there are epiphytal aroids and orchids, so there are also their prototypes on the sand-reef, in some cases differing considerably, and in others but little. The great Anthurium of the tree-top is represented on the sand-reef by a similar species, and one form of the Catasetum tridentatum that grows on branches beside the creek as closely resembles the Catasetum discolor of the mourie. Excepting some of the Clusias and figs, however, they generally differ so much as to be ranked as distinct species, showing considerable progress since the remote time when the separation took place. Of the second class the species of Pæpalanthus, the Schizæas, and a host of other dwarf wiry plants have no congeners in the forest—on the mourie or savannah they originated, and there they remain, every generation becoming more and more fitted to their habitat. The most striking examples here, as everywhere else are the orchids. We have already mentioned the Crytopodiums, but the Sobralias show even more adaptability to circumstances. So distinct are they, that a novice would put them down at once as dwarf bamboos rather than orchids. Instead of pseudo-bulbs, fleshy stems and delicate leaves, they have wiry stalks and dry hairy foliage, only the magnificent blossoms indicating at once their family. These have obviously been developed on the sand-reef or mountain, probably from something like the Cleistes rosea of the pipe-clay savannah. Another genus which must also have arisen in the same localities is the Vanilla, the only true climber in the order. Here we have an elongated fleshy stem either growing in the sand or entirely detached, and therefore an epiphyte as much as most of its cousins. By means of its power of storing moisture in fleshy aerial roots, stems, and leaves, it is able to endure the strongest sunlight and thrive where the others would be quickly shrivelled up.

In the forest proper all the plants are perennial

in duration, but on the mourie we begin to find a few annuals. We have already seen that development depends a great deal on the number of generations, and shown what a powerful influence towards that end is the flood. Now we have to deal with the drought, the influence of which tends towards the same result. If great trees on the creeks, with their burdens of creepers, epiphytes and parasites, are destroyed by the deluge, the three months of the long dry season play similar havoc on the vegetation of the sand-reef. It is, therefore, of the utmost importance that the plants should flower and perfect their seeds as quickly as possible, so that if the individual dies there may be always plenty of his offspring ready to come up when a more congenial season arrives. This accounts for the many flowers, their attendant insects and resulting seeds which we see here. Having to attain this object, they hardly care to waste the little plant food they can obtain on great stems and dense masses of foliage, but rather devote their energies to the production of flowers. Few are so highly developed as the Vanilla and other orchids, which make such elaborate preparations for this contingency by storing food, so they have to make other arrangements, and these take the form of a provision in case of death. The species, or long chain of individuals with its myriad links, takes little account of any one of these items, as long as the line is continued—whether it lives a thousand years or only a few weeks matters little. It has therefore followed that the individual is sacrificed that the species may live, and many plants of the sand-reef have become virtually or actually annuals, in the latter case going through their different stages in one rainy season.

In the south-west of British Guiana is a large tract of country where the bed-rock is sand-stone, and this resembles the Mourie in many respects. The undulated plains which here break the continuity of the forest for large areas are called eppelings. Above the soft sand-stone comes a surface of hard caking mud, or in some places conglomerate, when unbroken resembling a pavement of beaten earth, generally of a ruddy brown colour. This covering extends over hill and valley, as well as over the undulations of the savannahs, and would make them real deserts if the rainfall were not so great. However, alternations of burning sunlight and deluge make great seams in the crust, into which water penetrates, producing the effect of a badly laid pavement of irregular flag stones without cement, here and there piled irregularly and alternating with pools of various sizes and shapes. Where unbroken the surface is absolutely bare, but as the cracks and pools are numerous, this barrenness is relieved by a fair sprinkling of vegetation. Naturally the plants of this region have to provide, like those of the mourie, for the long dry season, and they do so in a similar way. Here and there are clumps of bushes, sedges in the pools, orchids, and everywhere the widespread Pæpalanthus, which is such a stumbling-block in walking over these places as it is always in the way.

This sandstone region culminates in that curious group of mountains of which Roraima is the most conspicuous. Below the great precipice, which towers aloft to the height of fifteen hundred feet, is a slope where vegetation is as rampant as in the forest, although of a different character. Being about five thousand feet above the sea-level, the flora of this slope naturally differs somewhat from that of the plain, but the great difference comes from the excessive amount of moisture. Rarely indeed does a day pass without rain, often accompanied by strong gales. With a very slight coating of soil on the barren sandstone, it therefore follows that tall trees are entirely absent, and even those lower ones which manage to insinuate their roots into crevices of the rock and between the great boulders which litter the surface, are gnarled and obviously weather-worn. As if crouching on the ground to avoid the tempest, they form perches and shelter for such a jungle as can only be possible under similar circumstances. With almost no dry season to provide against, development has gone to extremes in certain directions, but perhaps not so far as might have been the case if the temperature were higher. Sobralia liliastrum grows here in great thickets, accompanied by tree ferns, palms, the tropical representative of the well-known English brakes (Pteris aquilina) and a Rubus nearly allied to the familiar blackberry. Here also the continual dampness is very favourable to the growth of hepaticæ and mosses, which cover almost every branch, and hang down dripping with moisture. In some places there are bogs, in which grow pitcher plants (Heliamphora nutans), slipper orchids, Xyris and Abolboda, and in others the Bromelias, nearly always full of water, afford a congenial home for Utricularia Humboldtii, which spreads its runners from one to another until they cover them like a network.

The struggle for existence is as beautifully exemplified here as elsewhere. As in the forest proper, there are no annuals, and no times for rest. It follows, therefore, that the fight is continuous, and every plant has had to accommodate itself to a moist atmosphere that would destroy the inhabitants of the eppeling or the mourie. As those which live in the swamp are suited to the flood, so these on the mountain have managed to adapt themselves to the mist and driving rain without injury.

Yet they are not perfectly fitted to their surroundings, and never will be so. The winds still blow them down, and the rushing waters carry their trunks to choke up the creeks in the valleys. Death is necessary, and therefore it is here, as elsewhere, a factor to be reckoned with. It is the complement of life, and essential to progress.

X.

ON THE SEA-SHORE.

THE coast of Guiana is made up of the alluvion brought down by the great rivers in past ages, which by the continual work of trees and sedges and grasses has been raised to the surface of the water, so that it can now be brought under cultivation. The original coast line once coincided with the sand-reefs, which are at present at an average distance of about twenty miles from the shore—all between them and the ocean is the result of floods and vegetation. It has generally been supposed that this deposit is entirely due to floods, but that such is by no means entirely the case we shall endeavour to show. As in the forest, along the banks of the rivers and in the swamp, plants have had a great deal to do with the result as we see it, and are still working quietly but continually for the same ends.

As certain trees have been developed to contend with the flood, so others are fitted for the struggle with wind and wave. In some respects the contrivances are similar, but here on the coast special arrangements have grown up which seem to be the result of long experience. Throughout the tropics the same species are extending the foreshores and building up islands, but little notice has been taken of their work. Therefore, although neither Guiana nor South America are peculiar in this respect, yet they are such grand examples as to be well worthy of note. From the great delta of the Orinoco to that of the Amazon, the courida (Avicennia nitida) and mangrove (Rhizophora Manglier) have been utilising the floods and fighting the sea for ages, with the result that thousands of miles of land have been raised mainly by their efforts. It is true that mudbanks would have been formed without their aid, but every one knows how unstable these are, and how liable to continual shifting from tide and current.

For fifty miles or more from the coast of Guiana the sea is tinged by the thousands of tons of suspended matter brought down from mountain and forest by the great rivers. Part of this is a vegetable infusion, which on coming in contact with salt water is decomposed, and falls to the bottom as a flocculent precipitate; the remainder clay and fine sand, which are mostly deposited on the fore-

ABORTIVE AËRIAL ROOTS OF THE COURIDA.

shore. In addition to these there are floating islands, bushes and trunks of trees, all of which go to help in collecting the other deposits and binding them together. But even with all these the result would probably be only a series of mudbanks had not the courida and mangrove come to the rescue.

The courida, which was confounded with the mangrove by old writers (who represented its fruit as falling into the water and changing into barnacle geese), is perhaps unique in its contrivance for making islands and extending muddy shores. At one time it must have supported itself in the ooze by aerial roots, which probably, like those of the mangrove, grew out from the trunk, and extended outwards and downwards to form props or buttresses. That this was certainly the case is shown by excrescences on the trunk, which throw out extensions that hang downwards for two or three feet and then stop. Even did they reach the mud they are obviously too weak to be of much use, but when we appreciate the fact that they are now nothing less than miniature representations of what once were great branching extensions, we can easily understand their original purpose.

Looking closer, we can also observe the reason why the courida no longer requires such buttresses. Let us walk along the shore at low water and enter the broad fringe of trees, which stands up like an advanced guard, and extends without a natural break except the mouths of the rivers along the whole coast-line. Like the forest of the interior, it is dimly lighted, but, unlike it, it is almost entirely free from bush-ropes, epiphytes, and parasites. Now and again we come across a mangrove, Laguncularia, or Thespesia populnea, but the general effect is monotonous, as might be expected from the want of variety. Unlike the trees of the forest proper, the courida branches considerably all up the stem, and is more like a Lombardy poplar than a canopied giant of the forest. The leaves are thick, and of a dark green colour, rather narrow, and comparatively loose, as if formed to allow the wind free play. Then, again, the trunk, especially when young, is more or less flexible, the result of the whole being that the tree bends rather than breaks even when exposed to the roughest gale. Unlike so many other trees under similar conditions, it does not bend away from the direction of the wind; nor does it, like the cocoanut palm, have occasion to recover itself from such a position.[1] Offering little resistance, it is therefore

[1] When this palm is bent backwards by the wind it braces itself and turns towards it, the alternate backward and forward growth producing that wavy stem which is so characteristic, and which has been accepted by artists as its natural shape. See chapter xiii.

well fitted for a struggle with the air, but it is against the water that it has developed its highest powers.

Continuing our excursion in the courida bush, we notice that although walking on a soft ooze our feet hardly sink into it. Now and again a place is softer than the rest, and we can at once realise what would be our condition were the whole surface nothing but this same " putta-putta," as the negroes call it. Everywhere at the mouths of the crab holes we see it in heaps, and if we should be unfortunate enough to step into it, feel at once that it has no bottom. The crabs, however, find it congenial to their tastes, for they are here by thousands, their scarlet and dark blue bodies and long arms, with which they apparently beckon to each other, enlivening the otherwise dismal shade. Again we wonder why the ground is so springy underneath, and look down for the reason, but except that here and there are patches of upright pegs, like sticks thrust into the mud, there is nothing but an even surface. Putting a foot on one of these sticks, we find it firm but elastic—it will move back and forth but not downwards, and return again to its first position when the pressure is removed. We ask ourselves whatever can be the use of these things? They are not suckers neither are they offshoots from the roots—there are neither leaves nor leaf-buds on any one of them. Looking closely, we find that they resemble the aerial roots or branches of figs and Pandanus, but who ever heard of these growing upwards? These latter are props and buttresses to keep the trees straight and prevent their falling; is it possible that those beneath our feet are also intended to help the courida in a similar way? How absurd it appears as the thought flashes through our mind! We smile at the idea of a prop rising into the air with nothing to arrest its progress or serve as a *point d'appui*. Then, again, all these pegs are on about the same level, and rise but a few inches above the surface. Yet they may be considered from one standpoint as only buttresses of a more perfect type than those of the mangrove.

How can we prove this? Not by remaining in the depths of this great plantation, but by walking along the shore when the tide is low. In many places this is virtually impossible on account of the ooze, but here and there the stronger currents deposit banks of sand, on which we can walk without difficulty. Now we can see exactly how the roots of the courida are matted and interlaced one with another, and what a revelation it is! What a perfect breakwater this is—a fascine dam which

MATTED ROOTS OF COURIDA.

never rots, but goes on from day to day, becoming more and more perfect as the waves wrestle with it. To the naturalist the picture is one which brings the struggle most vividly before his eyes. For days and weeks, perhaps, this confused layer of roots has been covering itself with mud and silt, until hardly a fibre can be seen. Then comes the spring tide with a strong wind, which washes away almost everything, leaving it open to sun and wind. Any other tree would be at least scotched, if not killed, by such a catastrophe, but not the courida. He stands up as bold as before, confidently trusting that the great mat will always lie upon the mud whether it be covered or bare. The high tide will soon recede, and the work of gathering silt be renewed. It will be seen from the illustration how beautifully fitted is this living mat to collect and hold whatever floats in the water, but as if this were not enough, we have those curious peg-like aerial roots, which are a host in themselves. Where the roots have been bent and twisted they are naturally somewhat confused, but near still water they rise like a miniature forest of bare stems. This is beautifully illustrated in the view taken beside the draining canal of a plantation. Here there is little opportunity for collecting silt, and therefore the pegs are almost bare, but inside the courida bush they can never be seen in this state, but only peeping from the mud for an inch or two here and there. It can be easily understood that when every crevice between the tangle of roots is filled with mud and layers of the same are held together by the pegs, the tree is so weighted down as to be able to defy every wind that blows, and say to the waves, "Thus far shalt thou come, but no farther."

Notwithstanding all its precautions, however, its victory often recoils upon itself. Year after year it goes on extending farther and farther, reclaiming acres of mud flat from the raging waters, until perhaps a headland is produced which forms a great obstruction to some particular current. Then ensues a greater struggle than ever. The northeast wind comes at a time when the tide is at its highest, dashes the waves against the barrier, they undermine it, drag away one great mat of roots after another, and again flow past where was once the little cape. But, even then the courida has the best of it, for rarely does it happen that the sea recovers all it has lost and the general result is something added to the muddy fore-shore. Again, there are times when, notwithstanding all their exertions, the waves do not succeed in removing the barrier, and then they have to take themselves off to some other quarter where the dam is weaker. Perhaps the manager of a plantation has cut down his advanced guard to admit the pleasant breeze and cool his verandah. Here is an opportunity for vengeance, and soon the masses of dead roots are carried off wholesale, the mud dam is washed away, and the front of the estate flooded with sea water. When such an opening does not occur the current goes on looking for the weakest point in the long line. Woe betide the courida if it is not prepared; it is lifted up, carried away bodily, and dashed to pieces by the waves. Even yet, however, the courida may have kept guard so well that all the efforts of the raging current cannot find the smallest opening. Then it goes scouring along the coast, deepening a channel here, filling up another there, now taking a foot off a sand-bank which was dry at low water, and then throwing its tons of suspended matter aside to form the nucleus of an island, to the utter consternation of the pilots whose calculations are upset by the catastrophe.

Perhaps our readers may think we are going too far in ascribing all this to the work of the courida, but we can assure them that such things are continually happening here, and that they are primarily due to this wonderful tree admits of no dispute. How often an estates manager has had to rue the day on which he cut down the living barrier which obstructed his view of the sea, is hard to tell, but that many have suffered on this account is certain. It would not be difficult to collect the details and show how the front lands of sugar plantations have been flooded, the canes destroyed by salt water, and thousands of pounds spent in repairing the damage —all perhaps from thinning (and thus weakening) the living barrier.

We shall not, however, go into these details, but proceed to show how the courida builds up islands and how quickly this is done. A century ago the east coast of Demerara was drained by a creek which had its exit near what is now the sugar plantation Lusignan. At that time there were no estates in this district—the drainage was nature's work, and she did it in a way entirely suitable to the conditions then existing. A little later, however, one plantation arose after another, each with its drainage canals to carry off the water from its own area. It naturally followed that the Courabanna Creek became diminished in its volume of water as well as in its velocity, with the result that it could no longer contend with the banks of mud thrown up by the tide at its mouth. Soon an extensive bar was formed, and later this rose to the surface and became a mudbank, forcing the waters of the creek to pass round on either side. The eddy between these two streams allowed more and more silt to deposit until a fairly large bank was formed which became exposed at low water. Then came the courida, whose seeds were floating everywhere and had been washed over the bank times out of number without securing a foot-hold. Now, however, a few of them plunged their roots into the mud when the tide was slack for a few days, and held their own against the next springs. A commencement had been made, the seedlings grew to bushes, the bushes to trees, and soon a little grove was established, which continually extended itself as the island became larger. By and by it became more than a mile long, and the divided outlet of the creek was known under the two names of Great and Little Courabanna. These, of course, imply the fact that one was of more importance than the other, and as the weakest always goes to the wall, it was not long before the courida crossed over to the mainland and obliterated the smaller creek. Thus the island became a cape, and was known as Courabanna Point, or Point Spirit. As more plantations were laid out, less and less water came down the creek, obstructions of various kinds were formed, and soon after the beginning of this century the creek was obliterated, and is now entirely gone, leaving only what is called Lusignan Point to indicate its former site. Whether this little headland is on the exact spot where the coast pilot once recognised Courabanna Point is doubtful, as these headlands are continually changing in the struggle between wave and courida, but although perhaps the pilot might dispute its identity, it is as much the same as any other part of the coast can be after such work has been going on for a century, or even fifty years.

How important these changes are will be understood when we state that the twenty miles of coast between the Demerara River and the Mahaica Creek was once well known under the name of the Courabanna District. But it could easily happen that most important issues might depend on the existence or non-existence of such a creek. When the colony of Berbice was settled, the boundary line with Surinam was fixed at the Devil's Creek, about midway between the Berbice and Corentyne Rivers. That creek has gone the same way as the Courabanna, and it would be very difficult to indicate its mouth, much more its channel, at the present day. Fortunately, however, the Governor of Berbice in 1799, when both colonies were in possession of Great Britain, made an arrangement with his *confrère* in Surinam to make the great river Corentyne the boundary, and this arrangement having been retained when Surinam was restored to the Dutch, a great difficulty and possible occasion of a boundary dispute have been obviated. It is easy to conceive from this how important is the struggle between the sea and the courida if our readers were not convinced by the apparently wild statements made a few pages back.

However, we have not yet finished the story of this marvellous tree. At the beginning of this century the charts of the mouth of the River Essequebo showed a bank of "hard sand, dry at low water," to the east of Leguan Island. This place continued as a sandbank for over sixty years—how long it had been in existence before is doubtful, but we may safely state that it could hardly have been less than a century altogether, and from all appearances it might remain in the same condition for as long again. About the year 1862, however, an estates schooner, named the *Dauntless*, was wrecked on this Leguan Bank, partly broken up and embedded in the sand, where its presence was shown by a slight elevation, and one or two ribs sticking out above the surface. These jagged points arrested a few pieces of the tangle which came down the river, and on this were deposited some seed of the courida. Then began the work of building up an island which to-day is about two miles long by one broad, and is known on the chart as Dauntless Island.

When we see the magnificent results of Nature's operations our own feeble efforts seem reduced to almost nothingness. True, we may go to her school and learn a thousand lessons, by the carrying out of which we may attain some measure of success. But where can we find such a perfect example of combination as is shown by this grand line of sea defences. Even the dams and dykes of the Hollander sink into insignificance before it. But here, as in the forest proper, Nature's operations can only be thoroughly understood when carefully investigated from a thousand points of view, and when even what appear to us the merest trifles are taken into account.

Leaving the courida we must now enter the mouths of the river and admire the work of the mangrove. We have already stated that it is found here and there in the courida bush, but to investigate it properly we must go a few miles up one of the larger streams. Here the banks are almost entirely lined with its apparently inextricably confused jungle of buttresses. We can walk in the courida bush, but not in the mangrove swamp. Here is

IN A MANGO SWAMP.

no springy mat lying upon the mud, but every forking aerial root strikes downward, spreading a little it is true as it reaches the mud, but leaving almost bottomless mud holes between. The only possible means of getting through such a jungle is by crawling from one arched buttress to another at the risk of continual slips, and with a result which is only describable as being "up to your eyes in mud."

Unlike the courida, the mangrove does not attempt to keep back the water—it seems to know that any attempt of the kind would not only be rash, but sure to end in its destruction. When the floods come to meet the tide there is such a churning that it would seem impossible for anything to stand in the way, yet the mangrove remains securely at anchor. To do this, however, it has had to give up all the traditions and examples of other trees, and live without a body or trunk. True, there is a main stem, but it is very small in proportion to the spread of branches, and is hardly distinguishable from the larger buttresses which extend outward and downward like the long legs of a spider without its bloated abdomen. Like the trunk of the courida, these buttresses are somewhat elastic, and are thus able to stand the turmoil of a flood, when, if they were at all brittle, they would be broken in pieces. It may be easily seen that with this contrivance the mangrove hardly offers any opposition to the flood, but allows it to flow freely through its maze of roots. If a floating island should be entangled in its meshes when coming down the river, it is generally set free when the tide rises, as there are no thorns or protuberances of any kind upon the buttresses or branches. The result is that whether the water be high or low the tree remains quite comfortable, and lets the fishes play about its props to their hearts' content. As it does not rise to any great height it is equally safe in the strongest gale, as there is absolutely nothing to overthrow. Even if we could conceive of its being uprooted and carried away, the result would be simply transplanting to another spot.

Perhaps the most interesting thing about the mangrove is its fruit. The radicle commences to grow while the seed is still attached to the branch, and before it drops is generally a foot or more in length. Like great spike-nails, clubbed and pointed at the lower end, they hang straight downwards, ready when the time comes to drop and stick themselves upright in the mud. Then they sprout above and penetrate the mud with their roots, after which they quickly rise and throw out their characteristic buttresses. Here we have a suggestion of the pile so well known in connection with bridge building. However ancient the lake dwellers may have been, there can be little doubt that the mangrove learnt this little bit of engineering long before them. Then, in the cluster of sprawling buttresses we have the same contrivance which is used to support beacons, and here also the tree was first in the field. The contrivance of the courida is a fascine dam far more perfect than anything erected by man, as it cannot decay. When a man invents a thing of this kind we honour him for his ingenuity; what shall we say of the tree?

We must not leave the mangrove and courida swamp without calling attention to the fact that very few plants can endure salt water about their roots for even a few hours without serious injury. Besides the two species we have been considering, there are, however, several others all more or less fitted to the shore. The most conspicuous is the Laguncularia, which assists the courida, but is not so highly developed; after that we may mention the handsome Thespesia populnea with its hibiscus-like flowers. Beyond these the number that revel in salt water is small, and for this reason we do not get that variety in the courida bush which is so conspicuous in the forest proper. It is such a radical change in the economy of a vegetable to alter its assimilative arrangements in such a manner that it can be easily understood what enormous efforts must have been put forth before this took place. We believe that no great tree has succeeded in a similar way in temperate climes, although a few shrubs and herbaceous plants have been more fortunate.

XI.

IN THE TROPICAL GARDEN.

IN striking contrast with the plants of the forest those of our fields and gardens appear but poor weaklings. For centuries they have been more or less fostered with, at least from nature's point of view, very sad results. The forest giants have become strong and able to endure the vicissitudes of the weather, to hold their own in the struggle with others of the same species as well as a host of animal and vegetable enemies, while the more delicate fruit trees under cultivation can do nothing without the fostering hand of man. True, we bring together a host of beauties from a hundred different environments, and manage to keep them alive with a great deal of trouble and care, but they are never so interesting as when in their native habitats; and, although some thrive well, what a great number merely vegetate. They cannot endure the flood or drought, but must be watered, or the land on which they are grown well drained. Few have made any provision against the blood-sucking loranths or scale-insects, and consequently suffer much, hardly ever succeeding in throwing them off without assistance. Those which have been in cultivation for long periods are especially weak, while others coming from a different environment have never had the same pests to contend with, and are therefore unfitted to their new surroundings. The loranths seem to know this, and flourish in the utmost luxuriance among their branches, killing them very quickly if man does not interfere. The number of mango trees, for example, in the city of Georgetown and its vicinity which are always suffering from the attacks of a pretty stellate scale and a fungus like the well-known London "blacks" is enormous. Which comes first in the field is doubtful, but, speaking generally, we think the disease commences with a general weakness of the whole tree. Then probably comes the scale insect, fastening itself on the under side of the leaf, and increasing the debility until a state of general ill-health is produced. Finally the fungus covers the upper surface with its black film, which here and there peels off and hangs down in a most ragged and unsightly manner. The tree struggles to get rid of its tormentors, and perhaps, if it could

A JERMINALIA OVER-RUN BY LORANTH.

strip itself entirely for a few months, would succeed. But no; it can only drop its leaves a little sooner than usual, and bring out new ones to be quickly covered like the others.

The Barbados cherry, like the mango, is a foreigner in British Guiana. It is one of the strongest shrubs in our gardens, and is commonly used for hedges. So rampant is it in its growth that it bears cutting to almost any extent. You look at it closely, but can find no signs of weakness, and would expect it to be free from the attacks of the mango pest, and would be right. But everywhere about our gardens is the deadly loranth, and the cherry falls an easy victim from its very luxuriance. A dry, hard, and juiceless tree would be of little use to such a blood-sucker; it likes those that are fat. The more sap the longer the bush or tree can stand the continual depletion, and this is well exemplified now. As the loranth winds its aerial roots round branch after branch, and applies its sucking disks one after another, you see the little plant grow bigger and bigger, until it covers and obliterates its victim altogether. Meanwhile, however, to escape destruction, it takes firm hold of a neighbour and repeats the same operation with that also. It naturally follows that in many cases it comes to the ground, but meanwhile it has ripened thousands of fruit, which the birds have carried far and wide, so that its progeny is almost certain to be numerous.

The fiddle-wood tree (Citharexylon) is commonly cultivated for its deliciously-scented white flowers, the perfume of which is diffused for long distances in the evening. Like most white flowers, it is fertilised by the agency of a moth, which is attracted by the powerful odour in great numbers. Its leaves are delicate as compared with so many others that are either hard and dry, or thick and leathery, and are therefore more subject to the attacks of larvæ. It appears that the same moth which fertilises the flowers also deposits its eggs on the leaves, with the result that in a few days after these are hatched the tree is actually stripped bare. As a rule, however, it puts out all its energies, covers its branches with new foliage, and by the time the moth emerges is covered with flowers which afford nectar for its enemy or friend. Which of these two names the moth should bear is almost as difficult a question as any brought before a debating club, as in the larval state it is undoubtedly a pest, while later it is just as certainly an indispensable helper. Probably as a friend it does most good, for after all nothing is so essential to the continuity of the species as the production of seed.

Even if the tree should ultimately die of the strain produced by several strippings, it matters little from nature's point of view, although the loss of one of the line may spoil the appearance of an avenue.

In our gardens we have some very handsome so-called lilies. There is the Hippeastrum equestre, Crinums, Pancratiums, and the well-known Eucharis Amazonica. Some of these are well known in Europe, but there, even if they were subject to the attacks of larvæ, moths cannot get into the glass houses. Here, however, it is different, and the poor plants suffer much from their attacks. The perfect insect lays its eggs on the under surface of their leaves, and, if the gardener does not search for them, some fine (or wet) morning you see your beautiful plants turned into perfect wrecks. Hundreds of ugly worms are crawling over them, their leaves are nothing but thin membrane, and hang down in rags, while, if you have not discovered them within two or three days after they are hatched, it is almost certain that one or more have penetrated the bulb and are eating out its very heart. We have never seen destruction like this among these plants under natural conditions. Why, then, should it be so common in our gardens? This is a question which might be asked of several other plants, but hitherto we have been unable to find a satisfactory answer.

Nevertheless, from the very fact that we are not successful in preserving our gardens from the many pests which ravage them, they are all the more interesting to the naturalist. A greenhouse, where everything looks clean and tidy, may be more pleasant from one point of view, but if we want to understand the true meaning of the struggle for existence we must carefully observe what is going on in the tropical garden. When we shut up our beauties in glass houses, even if the insects and seeds of parasites are to be found outside, they cannot enter to disturb the plants. Here, however, in the open, every little difficulty and trouble is patent, and even when we do our very best to help them they often suffer.

A well-laid-out garden is no doubt very pleasant to the eyes, but after all there is always more or less of the artificial element introduced. We trim and prune until often every bush and tree grows in an unnatural shape, dwarfed, stunted, or lop-sided. Even the removal of parasites entails more or less deformity, as we cut away a branch because the loranths have taken possession of it, or spoil the shape of a palm to destroy the scale insects. Then, again, if we do not want our best specimens smothered by the more hardy and rampant trees, shrubs, and especially climbers, these latter must be hacked about here and there until they become quite deformed. This is the great fault of a tropical garden. When it is laid out with a number of pretty shrubs the owner never seems to think that many of these will grow to great trees, and in doing so smother the rest. At first the beds of crotons, hibiscus, and other foliage and flowering plants can be looked down upon as we walk in the garden; a few months hence they will be above our heads, and, if not trimmed, form a great thicket. Instead of rooting up the majority to make room, almost every one gets mutilated, with the result that hardly a single shrub or tree is anything but unsightly to a lover of nature. This is carried out almost everywhere, so that even public gardens offend the eye by a crowd of deformities. This is entirely apart from the display of bad taste in cutting shrubs into outlandish shapes—fortunately that is not very common—which shows simply the want of appreciation of the fact that there is as much real beauty in natural form as in colour.

To us the wild garden is far more beautiful and interesting. Here we see every one at work straining to get ahead, some losing ground as the others overtop them, but nevertheless doing their level best to exist and even thrive in the midst of the greatest difficulties. Here also every flower and fruit is seen at its best, and, if the plants are native to the country, many a rare insect and bird is attracted to the little bit of jungle. For, such will ultimately be the character of the place if left alone, only varying according to the species of plants first brought into it.

The first scene in the great struggle is the fight with weeds. Throughout the tropics every piece of cultivated land is infested with these. Unlike the more delicate garden plants, they are sturdy and strong, suited to many places and circumstances instead of one, and therefore better able to live in the garden than those you have planted. Only natives are able to contend successfully with them, and you will have to help the weaklings at the commencement. It will not be amiss, however, to watch the struggle before destroying them entirely, as they not only fight with your pets, but also among themselves. Some are armed with prickles and spines, others with down and bristles, while a few of the less unsightly are enabled to carry on the struggle by means of tough roots, rhizomes, and tubers, which endure alternations of wet and dry as well as defy the continual use of the hoe. They may be scotched but never killed by the gardener

—only the smothering of the trees can eradicate them entirely. The others live by perfecting their flowers and seeds very quickly, so that when the parent is destoyed the children are ready to come to the front at every opportunity. Like so many weeds all over the world, they have a thousand contrivances for scattering their seed. The Ruellia tuberosa generally comes to the front in dry weather, when it produces its handsome blue flowers and capsules. These capsules are so sensitive to moisture that when the rains fall they open suddenly and shoot the seeds to some distance. Other plants have hooked capsules, by means of which they cling to the fur of animals and are often carried away for miles ; while a third class provide plumes to their seeds, which enable them to float upon the wind. But, above everything else, the grasses are most rampant. One of the commonest in gardens is the bahama grass (Cynodon dactylon), which is almost the only kind suitable for lawns. This, however, is unable to endure much moisture, and therefore when the heavy rains fall, unless the soil is very light and well-drained, it suffers greatly. Then, it is gradually ousted by the Paspalum compressum, which spreads its leaves almost flat, and continually extends outward, smothering everything in its way. Even this species, however, can hardly endure sodden ground, although it changes its habit and grows upright in very wet weather, but has to give way to the monarch of the roadside and the estates dam, Paspalum conjugatum—the sour grass. This monster is at the top of the tree as far as endurance is concerned, although it is suppressed to a certain extent during a drought. Unlike the two others we have mentioned, which are eagerly devoured by horses and cattle, this is disliked even by the goat. In a congenial season it effaces every other weed, rising to a height of two or three feet, and extending even into the public roads. In the garden it indicates bad drainage, and will flourish at all times if undisturbed, but on the roadside it becomes much reduced during the dry season. Nevertheless, by means of its multitude of seeds and creeping habit, it again comes to the front in a few weeks after the rains, to be heartily abused by the stock-breeder as something not only useless, but offensive.

As the bushes and trees in your garden begin to spread they gradually shut out the light from everything below until not a single weed can come to anything, and now that they have succeeded in gaining their first victory they begin the fight with each other. Some are already seen to be the weaker, and these, having once lost their places, must ultimately die if left without assistance. Then come the creeping and scrambling vines. You thought to have a fine collection of bignonias and convolvuli, so that humming-birds, bees, and moths might be invited to the garden. Already this effect has been produced, and sitting down in the midst of the hum of insect life you think how pleasant is the scene. You take care not to disturb anything, but let the creepers wander at their own sweet will to bring these winged visitors. But the young trees do not find it quite so agreeable. A mighty scrambler has reached forward one extension after another until it has hung a flowery pall over one of them, and is now stretching out towards others. In vain does the tree push its young branches through this covering, they also are quickly hid, and if you do not come to its assistance it will ultimately die. If it is an ornamental flowering or fruit tree the effect of this exclusion of sunlight soon shows itself in the unhealthy appearance of the foliage, and the want of flowers with their consequent fruit.

If left alone these scramblers extend over the whole garden, covering everything, and ultimately killing all that come within their shade, except perhaps the palms, which push through and throw them off as they ascend. This you do not want, and must therefore give them a judicious trimming now and again. This can be done without mutilation or spoiling the effect of your interesting wild garden, which becomes a centre where birds build their nests in the trees, bats hang themselves under the dense climbers, marabuntas affix their bag-like homes in the branches, and a hundred species of ants are at work everywhere. Here also a few harmless snakes are sure to lurk, and with the lizards help to make the garden an epitome of almost everything, and a little world in itself. You may quietly study the fertilisation of flowers, and watch the whole process; see the ants attack and defeat great larvæ and cochroaches; observe the mantis spring forward from its apparently devout position and clutch the fly which has come within its reach; and feel as if you are one of these instead of looking down upon them all with scorn and contempt. The snakes, spiders, and larvæ, are no longer repulsive, but creatures most beautifully formed, and as much entitled to live as yourself.

XII.

MAN'S FOOTPRINTS.

PERHAPS the most striking peculiarity of the Guiana forest is its almost entire freedom from man's handiwork. Nevertheless, his footprints are visible everywhere, only requiring careful observation to enable us to recognise them and their significance. When South America was discovered the coast was inhabited by Indian tribes that are now either extinct or have gone away to the interior, but here and there the signs of their former presence are still visible. On the sand-reefs when once a clearing is made, years and even centuries may elapse before it is again incorporated with the jungle. Again, there are always a few plants which follow the steps of the red man and remain behind long after he has departed. The most common of these is the pine-apple, which grows in great clumps on the sand, and from its aggressive nature prevents anything weaker than itself from coming to the front. Whether this plant is ever truly wild is difficult to decide—as far as can be judged it is always a sign of man's presence at some former period. Ages ago perhaps an Indian in crossing the mourie threw the crown of a pine-apple beside the path after eating the fruit, and to-day impregnable clumps indicate here and there the tracks used long before the discovery of America.

On the little hills which rise here and there beside the creeks, and which we should judge likely places for Indian settlements if we did not know the people had migrated, the pine-apple is nearly always to be found if the soil is sandy and barren. With it also grows the handsome belladonna lily as well as the krattee (Nidularium Karatas), the fibres of which have always been used for hammock-ropes. More rarely the arrow-cane (Gynerium Saccharoides), and a few Caladiums are also seen, but as they require a more fertile soil, their presence is inconsistent with abandonment for any long period.

The first European settlers arrived about three centuries ago, bringing with them trees of the orange family, bananas, and sugar-cane. This last has never succeeded in propagating itself or in holding its own against the more energetic

AN INDIAN PATH.

native plants. The others, on the contrary, endure for a time according to circumstances, and are sometimes found existing in abandoned settlements for a long while after they have been deserted. Now and again we come upon the lime or citron growing in what a stranger would call the virgin forest but what we know to be "second growth." Once there was a clearing here, but man relinquished his efforts to keep nature in subjection, with the result that the forest has resumed its sway, and will ultimately smother every intruder. If the soil is wet this result will be produced all the sooner, as hardly any cultivated plants can endure flooding. In congenial situations, however, where the land is high and not too rich, they may live for a very long time, getting hardier as years roll on and becoming almost useless as fruit-bearers, only serving to indicate that at some former period an European settlement existed on the spot.

Coming down to a later period, we find that useful plant the bamboo introduced from the East, followed by the mango, bread-fruit, and a few others. At that time the plantations were established to the distance of about a hundred miles up the principal rivers, and almost the whole façade on either bank was cleared of forest. Then the

Dutch colonist had his orchard near the house, and very often a line of bamboos along the shore. When, however, it was found that the alluvion of the coast was the most fertile, a general exodus took place, and nearly all the river plantations were abandoned. In a very few places some of the mango trees still survive, but most of the other fruit-trees are entirely gone. Still, when even these have been destroyed by the stronger vegetation of the forest, a clump or even line of bamboos still holds its own, and plainly indicates that once upon a time a sugar plantation existed on the spot.

Perhaps the traveller has seen its name and that of the owner on some chart of the last century, and wishes to find out whether there are any remains of the buildings. If it has been abandoned a long time—say fifty years or more—the forest is fairly open after an entrance is made through the thick jungle which lines the river bank; but if, on the contrary, man has departed but a little while, it can only be penetrated by the use of the cutlass at every step. To find anything in either case is almost impossible, even if the brick foundations of the mill or the great oven have not been carried off, and when the explorer is particularly fortunate he finds only a distorted heap, held together by the roots of figs and creepers. These have insinuated themselves between every crevice of the brickwork until hardly two remain cemented together, and then only prevent their entire separation by the hold they have on them by their roots. If he is fortunate the traveller may find the family burying-place, and here again the insidious work of the fig is even more apparent. No one has taken away the bricks or slabs to make pillars for their houses, as they may have done in the other case, and therefore all the confusion apparent here is entirely the work of the plants. Marble tablets are heaved up in every direction, cracked across, pushed aside, and some quite covered with the network of roots. What was once a regular, oblong cavity is now a hollow surrounded by uneven banks of root-cemented bricks, in which no trace of either coffin or skeleton can be seen. The wood-ants have been at work breaking down everything into mould, and what was once the "Edele Achtbarr Herr" has been greedily taken up by the great masses of fibres which cover and interlace everything.

Coming now to places that have been very recently abandoned, we see Nature actually at work obliterating the marks of man's presence. Rampant creepers extend from the forest into the clearing and cover the fruit-trees, weeds choke every shrub in the flower garden, the cissus (C. sicyoides) climbs all over the wooden walls of the building, and figs insinuate their roots between the brickwork pillars. Cut an opening through the mound of vegetation which covers what was once the dwelling-house, and you will see in the interior a thicket of roots which have pushed themselves through the roof and penetrated one floor after another until they reached the ground, where they spread and dispute every inch with their neighbours. The painted boards may appear almost intact except for cracks and crevices, but if you press your hand against them ever so lightly they give way. Even the finger can be pushed through with hardly the risk of a splinter, as only a paper-like film remains of the once inch board, the remainder being nothing more than an inextricable confusion of galleries with similar partitions excavated by wood-ants. The floor is, of course, in the same condition, and even the beams are little better, although they may perhaps have a small core of solid timber. A few months later a heavy rainfall will make the whole collapse and gradually sink to the ground, the figs meanwhile rising higher and higher on the rich food provided by the decomposing material.

The benab of the Indian and hut of the boviander (river negro or half-breed) are, of course, more quickly broken down, as they are so much more fragile than the houses of European settlers. Perhaps a gourd vine climbs upon the roof, and brings it down in a month or two by its weight and the action of the moisture that trickles from its leaves on to the thatch. What was once the cassava or sweet potato field quickly becomes a jungle of prickly solanums interlaced by a thousand creepers, and the whole clearing is an example of rampant vegetation and the intense struggle for the mastery of one by another.

It will be seen from this that although man's footprints may be discovered in the forest, they are by no means obvious to a casual observer. On the savannahs and sand-reefs, however, as we have seen, they are more permanent and conspicuous. The great plain of Pirara—the traditional site of the Golden City of Manoa, the seat of El Dorado—is covered with slight elevations. On these may sometimes be discerned the marks of footpaths which once led to the Indian settlements on the top. How long they have been deserted it is impossible to guess—perhaps for centuries; yet from a distance they are quite obvious, although not so noticeable when under foot. Here and there, too, are streamlets with stepping-stones from which now obliterated trails once led, and these also tell of a former population.

These little footprints of past generations are, after all, very insignificant, and do almost nothing to alter the appearance of utter desertion. With the exception of a few pictures on the rocks near certain waterfalls, there are no antiquities, and therefore there can be nothing of that human interest so conspicuous in the old world. Yet the country, as well as the people, must be very old, as can be easily proved by the cultivated plants. Like the food vegetables of the other hemisphere, the cassava, sweet potato, maize, and capsicum are never found truly wild, and show by the number of their varieties that they must have been under man's care and control for ages. Whether any of these could exist for a length of time away from his influence is very doubtful, but it is strange that the same thing cannot be said of the pine-apple. Here we have a fruit, obviously also domesticated, which nevertheless thrives everywhere on the barren sand in utter neglect. This seems to show that it is possible for a cultivated plant to stand alone, apart from man's care and attention, but it is quite obvious that few others are possessed of this capability. In most cases we should say that these would quickly perish from off the earth if left to themselves.

There is, however, a class of plants which indicate man's presence in the forest, although they are never planted by him. Wherever a clearing is made, no matter how far it may be from others, certain species come to the front that are never to be found in the virgin forest. Among them may be mentioned the silver fern (Gymnogramma calomelanos), certain thorny solanums, grasses, and scrambling vines. These, like the weeds of more populous countries, follow man wherever he goes, but do not spread outside the forest region to any great extent. In presence of the great army of cosmopolitan weeds of the tropics they are obliged to give way, but in their own particular environment they flourish to perfection. It sometimes appears as if their seeds have lain dormant in the earth for indefinite periods—perhaps centuries—and are only brought to the surface when the soil is disturbed, otherwise it is hard to say how they came into the new clearing. An orchid (Catasetum discolor) is found in great numbers about old charcoal pits, and here we have another example of the result of turning up the soil. Similar behaviour has been noticed in certain wild flowers which grow in the woods of

Europe: they can be found in young plantations, disappear as the trees become more crowded, and again come to the front after a clearance has been made. In such cases the time during which they lie dormant cannot be very great, but here, in many cases, it must be something enormous.

In the absence of other indications of man's former presence, it is to the cultivated plants we must look; and here we meet with a great difficulty at the commencement. Except the cassava and capsicum, there are no others in general cultivation by the red man, and these are utterly unable to contend with the wildings of the forest. The cassava is very quickly overcome, and although perhaps the bird-pepper may endure a little longer, the annual capsicums, in all their varieties, are even more weak. We are therefore reduced to the pine-apple, the krattee, the arrow-cane, and a few beenas, and of these only the two first are able to exist alone for any length of time. These, however, are so common, and are of such great evidential value, that we need hardly ever entertain a doubt of the existence of a settlement wherever they are found. A less obvious but certain indication is the tobacco, which almost always comes up when the soil of such places is disturbed, the seeds having lain dormant for an indefinite period.

Here in tropical America the cultivated plants are, as we said before, very few. Maize has apparently been introduced into Guiana, and is hardly known among the Indians of the far interior. Cassava takes the place of corn. Its root is boiled as a vegetable; grated and pressed, it is made into flat cakes, and the inspissated juice is used with capsicums as a sauce for meat. For a change sweet potatoes and yams are grown to a very slight extent, but the staff of life is always the cassava. This deadly poison must have been in cultivation for ages; it has protected itself against every wild animal, leaving man to make the discovery that all its noxious properties can be dissipated by proper cooking. In other climes edible fruits and vegetables of many kinds have been developed by cultivation; the Indian has virtually but this one. Yet he has succeeded in doing a great deal with it, and in its absence would starve. When, therefore, we look upon its beautiful leaves and luxuriant stems, we cannot but wonder how, in some bygone age, the way to eliminate its poison and utilise it in such different ways was first discovered. Nowhere, perhaps, in the vegetable kingdom is a plant more noxious, and certainly none so useful, yet the man of the forest must have found this out very early indeed. Here, then, we have evidence of reasoning power of a very high order, unless we credit the grand discovery to accident. The lower animals develop the power of assimilating poisons without injury; man drives them off by cooking.

XIII.

THE SENSES OF PLANTS.

WHERE plants can be kept under subjection as they are in temperate regions men are apt to think of them as almost inanimate, and to use the word vegetate as if it meant little more than a slow mechanical process. Here in the tropics, however, trees refuse to be treated with contempt—they are sentient beings very much alive to the circumstances of their surroundings. It is all very well to ascribe this energy to the action and reaction of temperature, sunlight, and rainfall, but if the multitude of variations were produced by these agencies alone we should expect something like uniformity instead of the almost bewildering variations of the forest creatures. Here, if anywhere, the atmospheric conditions are uniform, at least for any given area, yet the differences between species, and even individuals, are pronounced in the highest degree. If some particular end were generally desired it might be supposed that every species would endeavour to reach it in the same way, but instead of this, hardly two have attained their object by other than most diverse means. These multifarious results undoubtedly go to prove that plants like animals are not altogether the creatures of circumstance.

Reflex action is often given as the cause of many variations, but a purely mechanical process is decidedly inadmissible. When a tree braces itself against the winds, waves, and floods by spreading its mat of roots on the mud, throwing out props or radiating a circle of buttresses near the ground, these contrivances may possibly be thought the result of reflex action, but to us they appear to be much more than this. If it were not so, then why should every species choose a more or less distinct means to the same end. In the forest we have here a tree with doubly-compound leaves made up of tiny leaflets, and immediately adjoining another with great digitate expansions as large as umbrellas. Here is a climber which coils serpent-like round its support, there another swarming upwards by holding on with claws like a cat, further on a third ascending by means of adhering aerial roots, and near by others swarming up as boys get through a tangle of branches by pushing through and then spreading out their arms. All are exposed to the same influences and have responded in a different way.

Instead of mechanical uniformity we have even greater diversity than among men, probably because trees want the means of combination. Every one stands up for itself entirely regardless of its neighbours except to push them away when they become too aggressive. We call this selfishness, but can hardly conceive of its existence apart from a knowledge of what is to be gained by the struggle. A blundering fool can never be credited with such a vice (or virtue ?), but a calculating business man is often charged with it. When a man gains some particular object for which he has been long striving, we call him persevering, energetic, and industrious, and when a tree does the same we can hardly do less than give it due credit.

It is hardly necessary to state that plants are sensitive, or to attribute the faculty only to those popularly so called. Of the five senses credited to animals they certainly possess three—feeling, taste, and smell. True they have neither eyes nor ears, but men and women are not necessarily wanting in intelligence because they are blind or deaf. Like human beings when deprived of both these organs, trees enjoy the sunlight, and are even affected by the vibrations of loud noises. The real foundation of all the senses is undoubtedly that of feeling—all the others may be wanting without loss of communication with things outside, but when feeling is gone the case is indeed a sad one. That feeling is possessed more or less by all plants is certain, and in some this sense is developed to a higher degree than in many of the lower animals.

One of the most striking examples of feeling is to be seen in the cocoanut palm, and we put this first because the sense is not so appreciable in the trunk of any other tree. Leaves and flowers can be easily recognised as sensitive, but a palm stem might be thought almost inert. All who have been in the tropics will have noticed the beautiful *straight* columns of the palm trees—natural pillars that seem fitted to support the forest canopy. Whether thin as walking sticks or massive as the pillars of a house, they are almost invariably erect, with one grand exception—the cocoanut. Even the artist has grasped this fact, and always takes care to depict this particular palm with a graceful wavy stem. As a matter of fact the bends are rarely graceful; they have been formed for use not ornament, and tell their own story to all who are prepared to listen or rather observe their doings. They are by no means all alike—one will lean in some direction with one or two almost zigzag bends, another is inclined at a different angle and nearly straight, while a third approaches the artist's ideal, but never comes up to it. Along the shore there is generally more uniformity; the stems approximate to the shape of a bow, the two points of which, the bole and the head, point seaward. They behave in fact exactly in the same way as men and women under similar circumstances, facing the wind and bending forward to keep themselves steady.

While other palms belong to the forest, the cocoanut, on the contrary, is a sea-shore plant all over the tropics, and has become suited to its environment. But it is by no means at a standstill; on the contrary, as we have already stated, the trunk is particularly sensitive. As the young palm commences to rise we see its stem facing the wind and curving forward, but when taller and exposed to the full power of the continuous and sometimes excessive breeze, it is liable to be twisted to either side. Then we can also see how it struggles to recover its first position, bending forward at the head and growing a foot or two until it is again blown backwards a little, to renew the struggle as before. The result is that, with the many differences of locality and consequent variations in exposure, every stem is differently curved, twisted, and bent, some leaning in this direction and others in that, the only point at all general being towards the sea, and against the prevailing wind.

Other trees do not show themselves to be equally sensitive in the trunk, but there are many signs by which they can be recognised as not devoid of feeling. It is, however, in the roots where this sense predominates, but we need not quote examples as they will occur to every naturalist. Those plants called "sensitive" are equally well known, as are also the fly-catching sun-dews. They may be adduced to prove that leaves do many things that are not explicable by mere reflex action. As for flowers we have already dealt with them to a certain extent, and will only mention the well-known fact that the stamens are often peculiarly irritable, which is the highest development of feeling.

Taste and smell are so intimately connected with each other, as well as feeling, that they can hardly be considered apart. Roots are undoubtedly able to distinguish suitable from unsuitable food, and though they may be poisoned now and then, this is nothing strange as the same thing happens even to man. Their sensitive tips go wandering in every direction, branching here and there in search of proper food. As long as the soil is uncongenial they press forward, and only when a good feast is discovered do they throw out that broom-like mass of fibres so conspicuous on the banks of rivers and creeks. A barren subsoil is carefully avoided by keeping to the surface, while in the rich river bottom the sour water-logged alluvion is equally distasteful. On the sand-reef the tap roots go down fifty feet or more and spread most evenly to glean every particle of food contained in the water which has percolated to these depths. On the mountain, again, every chink and cranny between the rocks is explored, the roots sometimes penetrating through narrow crevices into hollows where water has accumulated and spreading their network of fibres over the roof, down the walls, and into the pools. In some cases it appears as if the roots smell the water at a distance, and move straight onwards until they reach it. Some epiphytes that push their aerial roots down the trunks of trees in the forest, hang them quite free when above the water, only allowing them to branch out when they reach the surface. In the first case, moisture is obtained from the rain and dew as they trickle down the little channels in the bark, while in the other a reservoir of water is below, and the plant seems to know it.

It is however in the aerial roots of orchids that sensitiveness culminates. On living trees, round logs, the bars of wooden baskets, and the sides of porous flower-pots, they flow along as it were and adhere so closely that it is impossible to remove them without injury. The fact that their support is congenial is so patent as to admit of no doubt whatever. Let, however, the bark of the tree become diseased, or the log begin to rot and the change is at once noticeable. The fungoid growths which develop under such circumstances are either poisonous in themselves or indicate a condition of things certainly deleterious to the orchid. The aerial roots lose their vitality, unfasten themselves from their perch, and sometimes prefer to fall rather than remain exposed to contact with anything so distasteful. In our garden we have seen a plant loose itself entirely, and have had to pick it up from the ground and fasten it to something more congenial. Another example we have been watching for several months—a plant of Burlingtonia attached to a thin board by means of copper wire. It has been established for about ten years, and seems in very good health, flowering at regular times, and adhering to its support until very lately by a good number of aerial roots. At present the board is hung against the trunk of a small tree, its back being in actual contact. To all appearance it is yet fairly sound although decay has probably set in, and it will be rotten in a year or two. Three months ago we noticed that the orchid was hanging over as if detached, and on examination found it had loosened itself as far as it could and was only supported by the wire. Looking closer again we discovered that it had thrown out several new roots which clasped the tree, and was prepared to swing itself from the one support to the other, had not the wire kept it back. In the forest no doubt such migrations are common—the branch becomes rotten, the orchid throws out extensions in every direction to search for something better, some of these attach themselves to a living bough, perhaps several feet distant, and the whole plant is moving off before its weak perch gives way. What is specially noticeable is the fact that it does not wait for the downfall, but "takes time by the forelock." In the case of the Burlingtonia it is noticeable that the aerial root which encircles the tree is much longer than any other that it produced while in its old position, and is evidently thus developed for a special purpose.

In many orchids, especially those of a large size,

there are always being produced a number of aerial roots which appear to be wandering about in search of something. They appear to twist to one side and then to another, now bending at an angle in any direction, and generally smelling out as it were for a congenial holdfast. Now they will run along a branch for a short distance and then, as if dissatisfied, raise their sensitive points and go farther or even free themselves entirely. Another time you will apparently see the point about to take hold, and a few hours afterwards find it moving off in another direction. All these things show that these plants are not only possessed of feeling, but almost certainly of taste and smell also.

Climbers are not wanting in the same faculties. They refuse to attach themselves to dead trees, and will have nothing to do with rotten sticks. Even the liking for one soil above another is an example of taste. Theoretically we like those things which are good for us—food that is suitable to our constitutions, elements which go to build up the tissues. That plants should have similar tastes is only to be expected; their sensitive root-points must have the power of choice, and that they make use of it is obvious.

In regard to smell, we know that certain gases are disliked by plants. They fold their leaves together, close the pores, and shrink, as it were, in abhorrence. Orchids can hardly endure a sea breeze, and never flourish unless protected from it. Other trees, like the cocoanut, barely exist when salt cannot be obtained, and seem to revel in what would kill the orchid. Every one knows that our most delicious perfumes come from leaves and flowers. Is it not probable that the plants themselves derive some satisfaction from them, even apart from their results in attracting insects? It would be a great anomaly to have a perfume distiller without the sense of smell, and we can see no reason why plants should not enjoy the odours they have taken so much care to manufacture.

Admitting that these senses are possessed by members of the vegetable kingdom, must we not conclude that they are discriminately used? Everything goes to prove that such is the case, and that, as far as lies in their power, they avoid what is disagreeable and rejoice in that which is pleasant. Every gardener recognises the signs of good health and vigour in his pets, and, on the other hand, observes the effects of disease and weakness. He sees them straining to overcome difficulties, and, when all chances of success are gone, make dying efforts to leave progeny behind.

This reminds us that plants suffer pain, which is only possible where there is feeling. Although they set to work and repair the damage done by a ruthless pruning, or that more cruel mutilation which produces a pollard, it cannot be otherwise than painful. Again, when the tree is in the folds of the strangler, or has its juices continually being abstracted by the loranth and scale insect, we recognise that it is suffering. Perhaps a great branch is torn off by the fall of another tree, and here we recognise the fact that it is wounded and bleeding. It tries its best to form a cicatrix, but this may be imperfect, or meanwhile the germs of fungi find a congenial spot on which to begin their work, which leads to mortification and ultimately death.

There are certain capabilities in some plants which can hardly be considered as evidences of feeling alone, but are yet intimately connected with it. They have latent powers that are never lost, although perhaps not utilised for many generations. Among these is the production of new plants on flower-stems and leaves of different species, and on the pseudo-bulbs of orchids. Among the plants of our gardens is one called the "tree of life" (Bryophyllum calycinum), which has the power of producing new plants on the edges of the leaves when these are cut off and hung up in almost any situation, or even when placed between the leaves of a book. This is by no means a general mode of reproduction in the species, nor does it take place under ordinary circumstances. We can even conceive it possible for centuries to elapse without its being utilised, yet the power remains, and can be easily proved by hanging up a leaf in the house. Under certain conditions the pseudo-bulbs and stems of orchids produce young plants, as do also the leaves of a few ferns. These may be considered as survivals from some former time when the plants propagated themselves in this way, and, as such, good examples of inherited capabilities, of which more will be said in the next chapter. Nevertheless sensation must be concerned in these revivals of ancient modes of reproduction, and no doubt they are the result of most delicate processes set in motion by feeling.

XIV.

THE CAUSES OF THE STRUGGLE.

WE cannot watch this intense struggle in the forest, on the bank of the creek, and beside the great waters without coming to certain conclusions. Here before us is strong evidence that every plant is straining after several things—light, room to grow, a secure anchorage against wind, wave, and flood, and, above everything, perfection in flower and seed. All have succeeded so far that they still exist, and, more than that, thrive to an extent far beyond what could be expected from a mere fortuitous concurrence of circumstances. They seem to know what they want, and obtain it to a greater or lesser degree, showing considerable latitude of choice in means to the same end. When a man does similar things we call him a reasonable being—why not say the same of an orchid?

It seems to us that all this is explicable on the basis that every species and every family is continuous as long as the line exists. Although, therefore, from one point of view, every living thing is an individual, in the wide sense it is only a link in the chain, and must be considered as part of one great whole, extending backwards to that misty period when life began on the earth.

We have seen that the Guiana forest is made up of hundreds of species instead of the half-a-dozen or so which crowd the woods of temperate climates. Not only is there great variety in this respect, but it may be safely affirmed that here also individuality attains its highest development. It is hardly necessary to state that every plant is an individual in the ordinary sense of the term, and as such differs more or less from every other.

In great cities the struggle for existence developes more individuality than in villages, where there is a tendency to perpetuate old manners and customs. A similar thing happens in the forest with a throng much denser than is possible in any city. We can only compare the trees to a great mob gathered in one place, and struggling with each other to gain the best points for observation. In such a crowd, although there is a certain amount of unison as far as the striving to attain the same object is concerned, nevertheless selfishness reigns supreme, and the weakest get trodden under foot. So is it in the forest; each individual strives to get a share of the sunlight, and elbows every one that stands in its way, in some cases climbing upon their shoulders, and in others trampling them down.

We can hardly conceive of selfishness without a self, and, if we go a little further, must admit that the word also implies consciousness. Few persons are able to appreciate the fact that trees are individuals, although few would question this in the case of such animals as dogs, cats, and horses, while the shepherd would claim distinctness for every sheep in his flock. Here in the tropics we have a class of plants—the epiphytal orchids—of which the individuals can be recognised from each other by every one who has studied them. Efforts are continually being made to arrange these into species and varieties, but every orchid fancier knows how difficult this is. No two are exactly alike. Apart from size of flower and leaf, which may be credited to the effect of surroundings, there are differences in colour, shape, and markings which stamp every one as peculiar and distinct. A careful observer knows his plants as he does his friends, although he cannot always explain how he does this any more than he could say why he recognises a hundred different negroes or Chinamen from each other.

Other plants are perhaps not so easily recognised as individuals, but there are characters even among the trees of the same species by which they may be known. Early and late flowering and fruiting individuals are seen in every country, often compelling attention by their differing from the average. Here such differences are carried to extremes, hardly two of the same species dropping their leaves, opening their flowers, or ripening their fruit at the same time. The natural consequence is that, instead of the majority carrying on these operations almost on the same day, the seasons last for a month or two, and it is possible to find in one garden or portion of forest individuals in a hundred different stages. This is not only a very interesting fact, but it has a bearing on the interdependence of animals and plants, which we have considered in another chapter, the result being a long extension of the flowering and fruiting seasons and consequent food supply.

Young animals are somewhat erratic; they have not yet got into a groove; their elders, on the contrary, are inclined to be more regular. The same thing occurs with plants; some trees which become quite denuded twice a year when mature are never bare in their youth. A similar thing takes place with their flowering; if a tree blossoms at very irregular times it is always a young sapling.

Other signs of individuality are recognised by the woodcutter, chemist, and tanner. Every timber dealer knows that there are great differences among woods, entirely apart from those produced by the localities on which the trees grow. These are popularly ascribed to the influence of the moon, and rules are laid down for felling only at particular stages of the planet's circuit. Every chemist and pharmacist also recognises the differences in the amount of the active ingredient in certain barks and leaves, while the leather preparer is as well acquainted with the varying amount of tannin in his materials. All these differences are undoubtedly due to the fact that some trees are better fitted than others for the great struggle for life, and have succeeded in secreting a larger quantity of the protective agent than their fellows.

More obvious to the popular eye are the variegations of ornamental foliage plants. Crotons are so well known as to have received the name of "match-me-not," from the impossibility of finding even two leaves on the same shrub exactly alike. Several species of Sida—common roadside weeds—are equally conspicuous in the same way, and in the forest variegation is by no means rare. As for shapes and sizes of leaves, although fairly uniform, yet a careful observer will notice minute differences in every one, and the same thing is more patent in flowers.

It is hardly necessary to say anything more to prove that individuality does exist among plants, as every gardener will be able to verify the fact from his personal experience. The great questions are, why does every one differ from the others, and how does this come about? This factor in evolution is not something belonging to a bygone time, but a real force continually at work. As individuals are born every day we ought to have no very great difficulty in discovering the why and how of these differences. Nevertheless, like other life problems, the causes of variation are still almost unknown, and it is doubtful whether we shall ever be able to obtain anything like certainty as to their origin.

However, even tentative theories are useful to the naturalist, and perhaps the views of an observer in the tropics may be considered interesting, even if they are not acceptable. We do not claim originality, in fact, our opinions are the result of study and research, modified probably by our own individuality. We have found them useful in many ways, and especially as means of arranging facts, which, without some connecting link, would be isolated, confused, and disordered.

We must premise that, accepting evolution without the slightest hesitation, we are met by the great problem of the origin of variation at the very commencement. The effects of environment when perpetuated become heredity, but this does not explain why two seeds, the offspring of the same parents, and developed under exactly similar circumstances, vary more or less from each other. Natural selection undoubtedly is a factor in the preservation of varieties, but when we come to it for an explanation of how they originate we are baffled. Henslow and Weissman have promulgated theories which, although not altogether satisfactory, are steps in the right direction, and students will be able to recognise their influence on our own, perhaps, more crude ideas.

The key to variation is undoubtedly to be found in sexual generation. Every one knows that varieties are more easily perpetuated by cuttings, suckers, and offsets, than from seeds. The sucker is not an individual in the same sense as the seedling, it resembles more those lower animalculæ which propagate themselves by division. These have been put down as deathless, for, as long as the species endures so long a part of the original individual, must live. A great leap was made when sexual generation was introduced. Instead of one parent with one set of experiences, two were brought together, possibly from more or less different surroundings, each impressed with the memories of their long existences. As a complement to this death followed, but when we look into the matter a little more closely we shall see that this development has not really altered the fact that as long as the species endures there is no extermination. Every new so-called individual is made up of a portion of each of its own parents, and strictly speaking, only differs from the bud or sucker in being an intimate compound of two instead of one. Every child is, therefore, not only itself, but its father and mother, as well as all their ancestors, and cannot be considered otherwise than a link in a long chain, which is only broken if he or she does not continue the line.

In the course of the thousands of generations through which a multiform but really identical personality has passed, its experiences must have been exceedingly numerous, and we can easily believe that every one of these has been perpetuated. In man this fact is continually being proven, and no doubt will be recognised by all when by means of photographs their ancestors can be properly studied. It has been said that "the face is the index to the mind," and however one may be inclined to dispute the inheritance or continuity of mind, he cannot say anything against the perpetuation of features. A particular shape of the nose, a curl of the lip, more or less deeply-set eyes, and a thousand little tricks and idiosyncrasies, are continually showing us that the father or mother lives again in the child. In some cases, where the likeness to the immediate parents is not obvious, it may be found in some remote ancestor, and if we only had their portraits, no doubt by careful study we should be able to trace almost every feature to its source. Even the palms of the hands and soles of the feet have peculiar lines which run in families, and go to make up a physiognomy, every part of which, if we could discover it, has a meaning.

Like a composite photograph, where one picture overlies another, and all go to make up the whole, the records of all ages are permanently fixed on every man and woman. It naturally follows that the latest impressions are strongest, and therefore children will be more like their fathers and mothers than more remote ancestors. Nevertheless, every now and again the latent characters come to the front, as they are always likely to do, and you can only trace them by looking for the likeness in the long gallery of family portraits.

If this continuity is so strikingly exemplified in the physiognomy, it can hardly be supposed to be wanting in the mental characteristics. And, here again we have an overwhelming mass of evidence to prove that virtues and vices are often exactly reproduced in the offspring. We commonly speak of a child as the image of his father, and include in that phrase not only a facial likeness, but often a peculiar temper or particular likes and dislikes. This is quite natural, the child could hardly be expected to have the outward signs without those inner feelings of which they are the exponents.

Going a step farther, it will not be difficult to understand that the results of the experience of all past generations is contained in every living thing, no matter whether they are conscious of the fact or not. Possibly each vital cell is impressed with the whole life history, and has a power of volition apart from the conscious will. When we act instinctively no doubt this power is brought into action, for instinct is nothing more than the result of past experience. Children do many things with good consequences where we know action has come from impulse and without knowledge. Again, the forest Indian and even the half-civilised negro does almost everything instinctively. They cannot give you a reason, for they have none, and when mothers ask their children why they have done so and so, rarely indeed can a rational answer be obtained. Even when reasons are given they are often so absurd that you feel inclined to class those who give them as half-witted rather than as rational beings. The fact is they do not reason, but simply carry on their daily work instinctively, or in imitation of some one else.

Presuming that the whole body is impressed with the experiences of all past generations, it can be easily understood that acting in unison these sensitive germs will often combine together and produce effects without the conscious exercise of the will. Again, the will may impress them to perform a certain action at a particular time in the future, and the thing will be done. The former case will perhaps help to explain the phenomena of dreams, and the latter the wonderful power possessed by certain persons of waking at a particular time. Then there is forgetfulness; perhaps you feel uneasy when the time has arrived for doing something and cannot remember what it is. It seems as if there are two memories, one which permeates the whole body and belongs to the continuous line of generations, and the other that of the individual life. The former is necessarily the strongest, as may be seen from the fact that our likes and dislikes are often unreasonable, and yet refuse to be governed by the will.

This physical memory or instinct seems to be common to both the animal and vegetable kingdom. It is a beautiful provision of nature for protecting every one of her creatures. On account of it young birds hide from the soaring hawk and the timid deer speeds away as man approaches, even when the particular individual may never have seen a bird of prey or a human being, much less suffered from their cruelty. In man it appears as a horror of snakes and other more or less noxious animals, probably inherited from ancestors far more subject to their attacks than later generations.

To return now to individuality; the more we study it the more difficult it becomes. Perhaps the very characters which we lay most stress upon are nothing more than the re-appearance of traits which belong to the family, and which have been obscured for centuries. Then there is another set of characters which have resulted from the environment of childhood and youth, and affect our estimate of the individual to a wonderful degree. In ordinary cases we might even go so far as to say that these account for almost everything. But, there are so many cases of genius and eminent capacity in certain persons, which neither heredity nor environment can explain, that true individuality may be considered proven. By this we mean, that even apart from the moulding of circumstances and the almost rigid fetters of continuity, every man differs from every other who is now living or has lived.

Animals and plants appear to exist mainly for the purpose of procreation. Everything else seems to be secondary to this. Some plants, it is true, thrive without producing seeds, appearing to concentrate their energies upon such inferior modes of reproduction as suckers, tubers, and rhizomes, but even these embrace every possible opportunity of flowering when they get enough light. All the preliminary work of the forest trees is done to find a place where procreation may be carried on, and when this is secured the principal task of their lifetime is accomplished. For this they build storehouses as it were and concentrate their strength for a supreme effort, the result of which we see in the lovely flowers which deck their canopies. In some smaller plants such as the orchids, we can appreciate what has to be done before a perfect seed is ripened. Many species have flowers so large and seed-vessels so heavy that we wonder how they could have been produced from such a little assemblage of pseudo-bulbs and leaves. Under cultivation many fail—they want to blossom but are too weak. And when they do succeed it is often by a last expiring effort—they have done their best to keep up the continuity, and then die. In some species of Catasetum there are male and female flowers, quite distinct in appearance the one from the other, and what is most curious, borne on the same plant at different times. It appears as if the orchid is able to choose whichever of these it is strong enough to bring to perfection. The male necessarily requires a less continuous strain as it is saved from the labour of ripening the capsules, which takes several months. When, therefore, the plant is comparatively weak, only male flowers are produced, but when the pseudo-bulbs are plump and strong, females. These and a thousand other examples of choice of means to a particular end, all go to prove that plants, like animals, are by no means wanting in what we have called physical memory and its complement physical reason.

Given a so-called individual, how does he, with the assistance of a mate, procreate another link in the chain that is in any way different from himself? The parents have inherited the experiences of all their progenitors and added to these a number of their own. They have also naturally settled down as it were and are little affected externally by change of circumstances. But it does not follow because we hide our feelings that, therefore, they do not affect us; on the contrary, is it not true that our passions may be stirred to their lowest depths until they dominate us entirely, when perhaps no sign is visible in the face? The sexual passion is undoubtedly the most intense throughout both the animal and vegetable kingdoms, and will therefore have more influence than any other. Even flowers are extremely sensitive at the time of fertilisation and some have been proved to generate heat during the process. This being the case it is not difficult to understand that the plastic germ of the future individual will become subject to impressions which would not affect either the leaf or trunk of its parents, and we believe these impressions go to make up that character which distinguishes the individual.

We may suppose a case where the flower has done its best to keep away noxious insects by means of certain secretions, and only partially succeeded. Some of the nectar has been ravished without any good result. The parent experiences this but is too old to change; it is the germ which is sufficiently impressionable to do this. After all it is only another experience of the continuous being, and differs little from the changes going on every day except that the infant is more plastic and more affected by the circumstance. We might go on quoting examples, but plenty of these will occur to the mind of the naturalist. Some will perhaps say that there are cases where no benefit can possibly arise from particular variations. We can never be quite sure that this objection is valid, and even were it the case it is easy to conceive that mistakes are made sometimes. It would be strange indeed if nature were faultless in this respect, seeing that man with all his instincts and higher development of reasoning power continues to err. And every mistake entails its own consequences which are often disastrous while natural selection perpetuates the most useful variations.

From these premises it appears to us that the key to variation is continuity. The experiences of every past generation is embodied in every living thing, and each one of these affects the offshoot more or less. To these are added, in the course of its own life, a thousand others, and when its most critical period arrives and the bud which goes to make up a new being is particularly sensitive, every little circumstance must then make so much the greater impression. When pain and trouble come man shows by his face that they are at work, however he may wish to hide them from his friends. How little a thing is pregnant with the direst results is seen in the cases of epidemic diseases, and it is a well-known fact that certain disorders of the sexual organs are very easily contracted because these organs are peculiarly sensitive at particular times. All these things go to prove that we have within us a host of memories of which we know nothing and that the record is being kept for all future generations. It is not for the naturalist to moralise upon such a subject, but nevertheless we cannot help remarking that, believing this, our responsibilities must appear overwhelming. No matter that everything is righted in time, much suffering is entailed in the process, and many generations may pass before the consequences of a mistake are effaced.

This reminds us of the fact that troubles and difficulties are essential to progress. Every variation and useful contrivance is the effort of imperfection to advance and become more fitted to a particular environment. But the end can never be altogether attained, as plants and animals both strive to overreach each other, and as fast as one succeeds in gaining a step in the way of defence, the other as quickly becomes more aggressive. The more difficulties encountered by the species the more experience it acquires, and therefore the greatest developments take place in those which have already overcome many obstacles. When a plant or animal has settled down as it were, and is no longer subject to the attacks of enemies or the vicissitudes of the weather, it begins to retrograde, as we have seen in the case of our cultivated flowers. It frequently happens, however, that this degradation is prevented by efforts in other directions—the plants do not require such hard, wiry stems and hairy leaves as when growing outside the garden, and therefore devote their energies to extraordinary developments in their flowers. Such cases, although from one point of view appearing to be weak and degenerate, are really examples of hard work and accommodation to new circumstances.

The theory of continuous existence seems to be very useful in explaining some of the problems of human life. Loti, in his "Book of Pity and Death," says: "The human head is filled with innumerable memories, heaped up pell-mell like the threads in a tangled skein. There are thousands and thousands of them hidden in obscure corners whence they will never come forth; the mysterious hand that moves and then puts them back seizes sometimes those which are most minute and most illusive, and brings them back for a moment into the light during those intervals of calm that precede and follow sleep. The commencement and the end (*of the dream*) existed in other brains long since returned to dust."

This is undoubtedly suggestive of that continuity which we have been considering, as is also the following from Tennyson's "Two Voices":—

> "Moreover something is or seems,
> That touches me with mystic gleams
> Like glimpses of forgotten dreams—
>
> Of something felt, like something here;
> Of something done, I know not where,
> Such as no language may declare."

THE END.

www.ingramcontent.com/pod-product-compliance
Lightning Source LLC
Chambersburg PA
CBHW032053220426
43664CB00008B/977